ments and Trends
g/10.5772/intechopen.1001645
Tu, Raed Abd-Alhameed and Ashwain Rayit

ditors: Beatrice Cheung

hmed Bouziane, Ankita Dixit, Benjamin Iñiguez, Dovranov Kuvondik Turakulovich, Hind
y Singh, Pedro Pereyra

London, United Kingdom, 2024 by IntechOpen
global imprint of INTECHOPEN LIMITED, registered in England and Wales,
r: 11086078, 167-169 Great Portland Street, London, W1W 5PF, United Kingdom

aloguing-in-Publication Data
for this book is available from the British Library

PDF copies can be obtained from orders@intechopen.com

ents and Trends
, Raed Abd-Alhameed and Ashwain Rayit

5466-647-8
-85466-646-1
78-0-85466-648-5

MOSFET - Deve
and Tren

Edited by Yuxia
Raed Abd-Alhameed and

Published in London, Uni

MOSFET - Develop
http://dx.doi.o
Edited by Yuxiang

Assistant to the

Contributors
Abdellah Aouaj, A
Jaafar, Indra Vija

First published in
IntechOpen is the
registration numb

British Library Cat
A catalogue recor

Additional hard an

MOSFET - Developm
Edited by Yuxiang T
p. cm.
Print ISBN 978-0-
Online ISBN 978-0
eBook (PDF) ISBN 9

For EU product safety concerns:
IN TECH d.o.o., Prolaz Marije Krucifikse Kozulić 3, 51000 Rijeka, Croatia,
info@intechopen.com or visit our website at intechopen.com.

Meet the editors

Dr. Yuxiang Tu is a researcher of electromagnetic and radio frequency engineering at the University of Bradford, UK. He was awarded a Ph.D. from the University of Bradford, UK, and became a teacher at a key university in China. He has published numerous international authoritative academic journal articles and conference papers. He has presided over and taken part in various research projects including the Nature Science Foundation projects in China, the European Union Horizon 2020 Research and Innovation Programme, EPSRC projects, and so on. He is also the reviewer of international authoritative academic journals and IEEE conference papers. His interests include wireless communications, microwave, radio frequency, and so on. He has been a member of the IEEE and the IET.

Dr. Raed A. Abd-Alhameed (M'02, SM'13) is a professor of electromagnetic and radio frequency engineering at the University of Bradford, UK. He is the leader of the Communications Research Group of the School of Engineering and Informatics, University of Bradford. He has published over 800 academic journal articles and conference papers; in addition, he has co-authored eight books and several book chapters including seven patents. He is a principal investigator and co-investigator for several funded applications for EPSRCs and so on. His interest is in computational methods and optimization, wireless and mobile communications, microwave, radio frequency, and so on. He is a fellow of the Institution of Engineering and Technology, a fellow of the Higher Education Academy, and a chartered engineer.

Dr. Ashwain Rayit was born in Bham, India. He received a Ph.D. in microwave engineering from the University of Bradford, UK, in 1993. Then he joined Filtronic Ltd. as an RF engineer and later became instrumental in setting up Filtronic Cable Communications Ltd. He developed a range of amplifiers and filtering products for the cable market. Later he set up SARAS Technology Limited with his brother. SARAS Technology is involved with the design and manufacture of a range of products (RF front ends and TR/TX modules, etc.) for many markets including defense, wireless communications, and security and surveillance. He has written numerous papers and contributed to a number of books. He has been a member of the IEEE and the IEE.

Contents

Preface

The book provides analysis and discussion about the developments and trends in metal semiconductor field effect transistors—MOSFETs. Renowned authors from academia describe novel MOSFET analyses and techniques in state-of-the-art solutions, enabling readers to deal with current high-performance MOSFET achievements and their promising future.

The book presents a comprehensive coverage platform of MOSFET to supply the reader with an integrative view of the latest research. The MOSFET is considered one of the most significant innovations in contemporary history, and it officially kicked off the third industrial revolution by greatly enhancing societal efficiency and productivity. MOSFETs cover a wide range of utilization in the modern academic, scientific, and industrial worlds. The related topics are numerous, such as MOSFET fundamentals, characteristics, operation, and design; MOSFET electronic circuits, systems, and networks; radio frequency engineering and microwave and wireless communication; MOSFET amplifiers, filters, transmitters, and receivers; and next-generation communication, 5G, 6G, communication standards, and so on. Hence, MOSFET analyses, techniques, and solutions are promising for large application in future MOSFET fields and play a vital role in society's advancement. This book will be useful for scholars, scientists, engineers, researchers, and students. It aims to combine both theoretical coverage and real-world application concepts. It will provide sufficient and valuable knowledge of various MOSFET applications from traditional to recent advances.

The developments and future of one of the most promising inventions—MOSFETs—are proposed in Chapter 1. In Chapter 2, the compact model of the DMG-GC-DOTTDCD cylindrical gate MOSFET is introduced and fully analyzed, which shows its superiority to its counterparts. The electrophysical properties of nanofilms that are generated using the magnetron sputtering method are well studied in Chapter 3, which is meaningful in MOSFET research fields. Chapter 4 outlines the novel theoretical approach for the analytical calculation of potential distributions and drift-diffusion currents in MOS and junctionless MOSFETs. The summary of the novel method that makes it possible to analytically solve the nonlinear Poisson equation in the inversion layer of MOS and MOSFETs is also presented. In Chapter 5, various device approaches including thin body, graded channel, halo doped, multigate, as well as source/drain are discussed. The functions of process parameters are carried out in the low moderate inversion region, which is meaningful in the fields of low power and high frequency.

The academic chief editor is Dr. Yuxiang Tu, who is from the University of Bradford. The co-editors are Professor Raed A. Abd-Alhameed, who is from the University of Bradford; Dr. Ashwain Rayit, the director of SARAS Technology Limited, and Ms. Beatrice Cheung, who is from the University of Manchester.

The editors would acknowledge the valuable comments by Dr. George Oguntala from Birmingham City University and the support of Ms. Tea Jelaca, Ms. Marijana Josipovic, and Ms. Jelena Germuth from the IntechOpen publishing house.

Yuxiang Tu
Faculty of Engineering and Informatics,
University of Bradford,
Bradford, UK

Dr. Raed Abd-Alhameed
University of Bradford,
Bradford, UK

Ashwain Rayit
Info SARAS Technology Limited,
Leeds, England

Chapter 1

MOSFET on the Horizon: What's New and What's Next

Ankita Dixit

Abstract

This chapter mainly enlighten about the development and trends of the field effect transistors (FETs) in the nanoelectronics industries. According to Moore's law, the number of transistors doubles in every 2 years because of transistor's size is scaled down. Though the scaling of MOSFET has been the driving force towards the technological advancement, but due to continuous scaling various secondary effect which include; short channel effects, high leakage current, excessive process variation and reliability issue degrades the device performance. In today's era, researchers are developing nano scaled transistors using various types of materials with different device geometries to reduce the limitations of conventional MOSFET. This chapter focuses on the development history, current-status and future trends of transistors. At the same time, the various protentional applications of nano-transistor discussed in this chapter.

Keywords: short channel effects, transistor miniaturization, gate-all- structure, field effect transistors, nanoelectronics

1. Introduction

In the ever-evolving landscape of semiconductor technology, the Metal-Oxide-Semiconductor Field-Effect-Transistors (MOSFETs) stands as a cornerstone, constantly evolving to meet the demands of modern electronic (Nano electronics) and shaping the future of the electronic devices. Gordon Moore, co-founder of INTEL, states that the number of transistors on a microchip double approximately every 2 years, leading to an exponential increase in computing power [1, 2]. The essence of Moore's law is the relentless miniaturization of transistors, allowing for more computing power in smaller spaces and at lower costs. This trend has led to the development of increasingly powerful and energy-efficient electronic devices, which are fundamental to the growth of the digital revolution.

This chapter delves into the dynamic realm of MOSFETs, exploring the latest advancements that define the present and peering into the exciting possibilities that lie ahead. As we embark on a journey through the cutting-edge developments and emerging trends, we unravel the MOSFETs on the horizon, shedding light on the innovations that promise to revolutionize the world of electronics. In this chapter we also navigate the current progress, uncovering what's new and contemplating the boundless potentials that awaits in the realm of MOSFET technology.

2. Development history of transistor

Figure 1 illustrates the evolution of the transistor technology. In 1947, the First working transistor (point contact transistor) was invented by John Bardeen, Walter Brattain and William Shockley at Bell Laboratory [3]. Shockley unveiled an enhanced iteration of the point contact transistor, known as the bipolar junction transistor (BJT), in 1948 [4]. This enhanced iteration, formally put into production in the early 1950s, marked a pivotal juncture that initiated the widespread adoption of transistors.

While the BJT brought about significant advancements, it also presented its own set of advantages and limitations. To address challenges such as low thermal stability, substantial noise and slow switching speeds, researchers sought the next evolution in transistor technology. Following the conceptualization of the Field Effect Transistor (FET) by J. E. Lilienfeld in 1925 [5], the first junction field effect transistor (JFET) was presented by Henrich Welker in 1945 [6]. Subsequently, in 1959, Mohamed Atalla and Dawon Kahng pioneered the silicon metal oxide semiconductor field effect transistor (Si-MOSFET) [7]. This marked a crucial milestone in transistor development, introducing a new paradigm with enhanced performance characteristics.

2.1 Bipolar junction transistor

As shown in **Figure 1**, Bipolar Junction Transistor (BJT) is a three-terminal semiconductor based current control device, which has emitter, base, and collector terminals. The main function of transistor is to control the flow of current between emitter and collector by using base terminal. Generally, the emitter current (I_E) is the sum of base current (I_B) and the collector current (I_C) [8].

$$I_E = I_C + I_B \tag{1}$$

Due to the limitations of the BJT as low thermal stability, low switching speed and high noise, researchers were looking into new electronic device [9, 10].

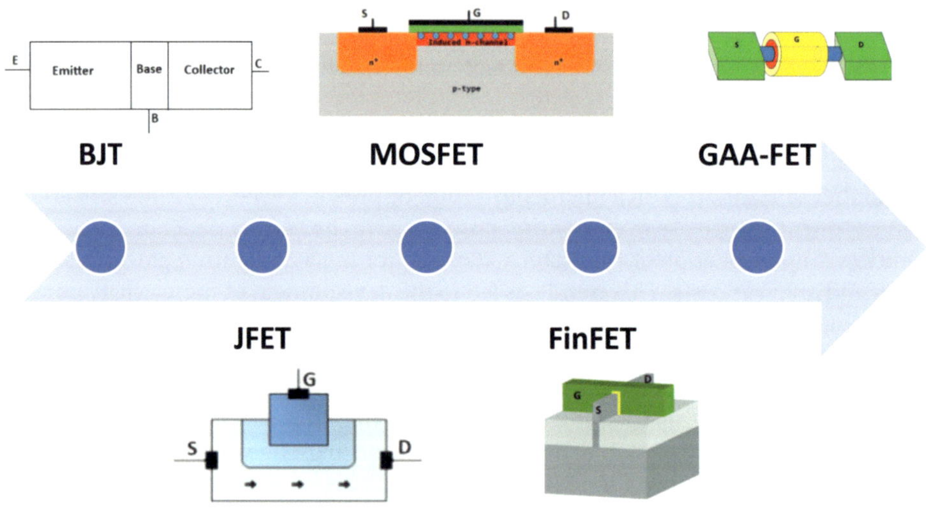

Figure 1.
Evolution of transistor.

2.2 Junction field effect transistors

Junction Field Effect Transistor (JFET) functions as a voltage-controlled transistor, characterized as a unipolar device featuring three terminals- source, drain and gate. The gate serves the purpose of regulating the output voltage. In contrast to Bipolar Junction Transistor (BJT), JFET exhibits superior attributes such as heightened thermal stability, reduced noise and compact size. However, its limitations in handling low signals paved the way for the emergence of Metal-Oxide-Semiconductor Field-Effect Transistor (MOSFET) [11, 12].

2.3 Metal-oxide semiconductor field effect transistor

MOSFET is another type of filed effect transistor which has a metal oxide semiconductor (MOS) structure. Similar to JFET, MOSFET has three terminals (source, drain and gate) which are used to control the conductivity between source and drain terminal based on the applied gate voltage. Based on the Moore's law, while the scaling of MOSFET has been the primary catalyst for technological advancement, the relentless scaling has given rise to several ancillary effects. These effects encompass short channel phenomena, increased leakage current, drain induced barrier lowering, and concerns related to device reliability, all of which collectively compromise the overall performance of the device [13–15]. The limitations associated with attempting to scale down traditional semiconductor devices have led researchers to look into the following possible solutions [16–18]:

- Alter the prevailing FET structure
- Employ a novel material as a substitute for silicon as the channel/substrate
- Utilization of materials with a high relative dielectric constant (ε_r, high for gate application

2.4 Fin field effect transistor

As shown in **Figure 1**, Fin FET is the three-dimension structure of a transistor. Instead of a flat channel, it features a fin like structure that rise above the substrate, providing better control over the flow of current. FinFET technology is commonly used in modern semiconductor manufacturing process, offering improved performance and energy efficiency compare to traditional transistors [19, 20]. It serves as the foundation for contemporary nanoelectronic semiconductor device manufacturing.

2.5 Gate-all-around field effect transistor

Gate-All-Around Field Effect Transistor (GAA-FET) is a type of transistor design where the gate material completely surrounds the channel region from all sides, providing better control over the flow of current. GAA-FET emerges as a promising advancement in semiconductor manufacturing, addressing scaling and performance limitations inherent in traditional transistor design [21]. This innovative architecture enables enhanced electrostatic control and mitigate leakage, surpassing the constraints of conventional transistor configurations [22].

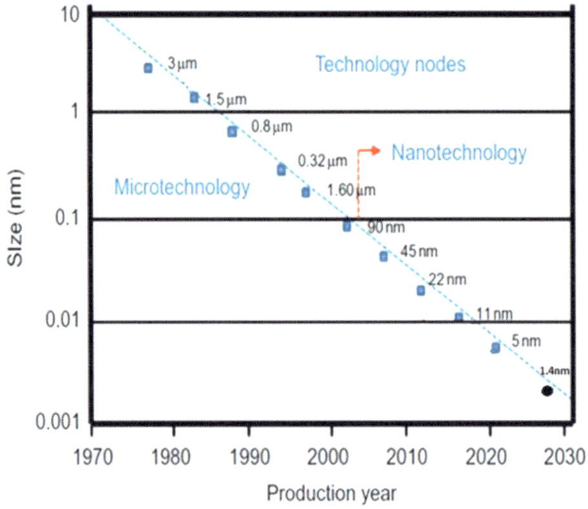

Figure 2.
Technology node roadmap for transistors.

3. Scaling

According to Moore, the principle of scaling involves increasing the density of transistor per unit volume on a singular integrated chip. **Figure 2** illustrates the scaled-down dimensions of the transistor across various technology nodes in semiconductor devices. Notably, the gate length is projected to reduce to approximately 3 nm by 2024 [23, 24]. The scaling of the Field Effect Transistors has been a driving catalyst for the progression of the nano-electronic devices. Nevertheless, persistent scaling introduces challenges such as short channel effects (including source/drain charge sharing, drain-induced barrier lowering, surface punchtrough, mobility degradation and velocity saturation) and heightened in leakage current [25–27].

The reduction in transistor size also leads to a decrease in oxide thickness, contributing to an escalation in gate leakage current. Consequently, for nano-scale design, the minimization of leakage power becomes equally imperative. The complications associated with scaling down traditional MOSFET prompted researchers to explore Gate-all-around structure-based transistors or nanotransistors. These novel transistors have undergone extensive scrutiny for diverse applications such as logic devices, memory devices, sensors etc. However, accurately predicting the ultimate performance of these innovative nanotransistors remain a challenge. To provide valuable insights, guide development efforts, and reduce costs, the development of precise and reliable simulation tools emerges as a paramount concern.

4. Nanotransistors

Semiconductor microelectronics has now transitioned into a genuine nano-electronic technology. Nanotechnology, defined by structures with at least two dimensions below 100 nm, material properties influenced by size and structures involving self-assembly, finds a perfect fit in silicon nanoelectronics [28, 29]. Semiconductor

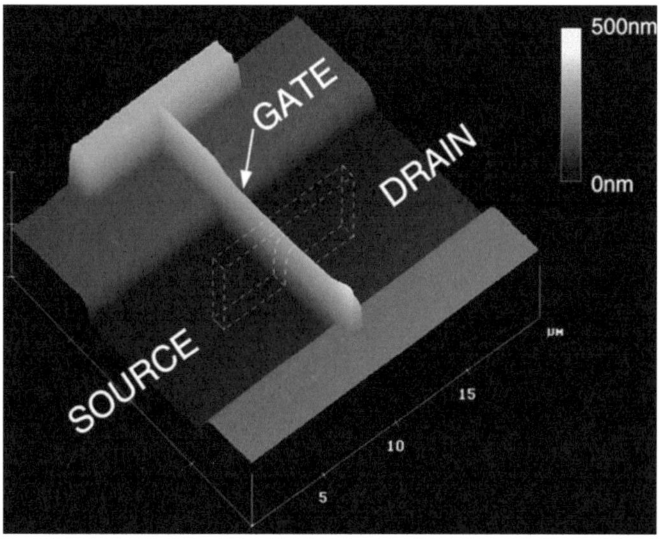

Figure 3.
A 3-D atomic force microscopic view of nanotransistor [30].

nanoelectronics full-fills all three criteria and can be recognized as the inaugural instance of active nanotechnology in manufacturing. As silicon technology undergoes the shift to nanoelectronics, a novel realm of research has emerged. Unconventional nanoelectronics leverages semiconductor nanotubes and nanowires, while molecular electronics employs molecules as electronic devices. The trajectory of this research towards practical electronic technologies remains uncertain, but its impact on practical nanoelectronics is evident in at least three aspects. Firstly, there is the development of new techniques for assembling nanostructures materials and nanoscale. Secondly, there is the advancement of novel methods for assembling nanostructured materials and nanoscale devices. Thirdly, there is a discernible comprehension of electronic devices at the nanoscale, called nanotransistors.

A transistor whose dimensions are in nanometers, called nanotransistor. **Figure 3** [30] shows the structure of nano-transistor. It is a compact and efficient electronic device and often associated with advancement in nanoelectronics and nanotechnology. The development of the nanotransistors is crucial for keeping up with Moore's Law, which trends has driven the miniaturization of electronic components, leading to more powerful and energy-efficient devices.

5. Applications

MOSFET find widespread applications across various electronic devices and systems due to their versatile and controllable nature. Some key applications are as follows (**Figure 4**) [31–41]:

1. *Amplifiers:* FETs are commonly used as amplifiers in electronics circuits. They can amplify signal with low distortion and noise, making them suitable for applications in audio amplifiers, radio frequency (RF) amplifiers and another signal processing circuits.

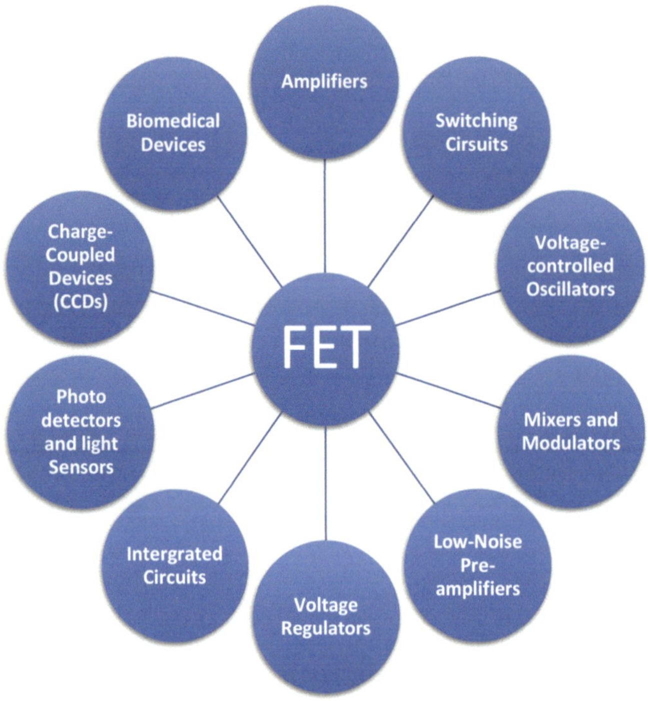

Figure 4.
Applications of field effect transistors [31–41].

2. *Switching Circuits:* FETs can operate as electronic switches in digital circuits. MOSFETs, are extensively used in digital integrated circuits, such as CMOS (complementary Metal-Oxide-Semiconductor) technology.

3. *Voltage-Controlled Oscillators (VCOs):* FETS can be used in voltag-controlled oscillators to generate variable frequency signals. This is useful in applications such as frequency modulation (FM) communication systems.

4. *Mixers and Modulators:* FETs are employed in mixer circuits used in RF and communication systems. They are also used in modulator circuits to modulate signals for various communication applications.

5. *Low-Noise Preamplifiers:* Due to their low noise characteristics, FETs are often used in the front-end of sensitive electronic systems, such as in low-noise amplifiers for communication receivers.

6. *Voltage Regulators:* FETs are used in voltage regulator circuits to regulate and stabilize the output voltage of power supplies.

7. *Integrated circuits (ICs):* FETs are fundamental component in the design of integrated circuits. Including microprocessors, memory chips and other digital and analog ICs.

8. *Sensors:* MOSFETs can be utilized as sensors in various application due to their sensitivity to change in certain physical quantities. Here are some ways in which MOSFETs are employed as sensors.

   Gas Sensors: MOSFETs can be used as gas sensors by incorporating a sensing material on the gate oxide. When exposed to specific gases, the electrical properties of the MOSFET change, leading to a measurable variation in the output current or voltage. This is commonly employed in gas detection systems.

   Chemical Sensors*:* MOSFETs can be serve as chemical sensors by modifying the gate oxide with a material that interacts with specific chemicals. Changes in the chemical environment result in a shift in the electrical characteristics of the MOSFET, allowing for the detection of a particular chemical compounds.

   Biological Sensors*:* MOSFETs modified with biomolecules or biological receptors can be used for sensing biological entities such as DNA, protein, or other biomolecules. The binding of target biomolecules to the modified MOSFET surface induced changes in the transistor's electrical properties, enabling the detection of biological interactions.

   pH Sensors: MOSFET can be configured as pH sensors by modifying the gate oxide with materials sensitive to change in pH levels. Variations in the pH of the surrounding environment result in changes in the MOSFET's threshold voltage or conductance, providing a means for pH measurement.

   Temperature Sensors: MOSFETs can be used as temperature sensors by exploiting the temperature-dependent characteristics of the transistor. Changes in temperature affect the MOSFET's threshold voltage and mobility, allowing it to function as a temperature-sensitive device.

   Radiation Sensors: MOSFETs can be sensitive to ionizing radiation. By using specially designed MOSFETs. They can be employed as radiation sensors in applications such as dosimetry and radiation monitoring.

   Pressure Sensors: MOSFETs can be integrated into pressure sensor devices by incorporating a flexible diaphragm that mechanically affects the transistor's electrical characteristics in response to change in pressure.

   Light Sensors: MOSFETs can be employed in light sensors or photodetector by utilizing their sensitivity to light. Light falling on the MOSFET's surface generates charge carriers, affecting the transistor's conductivity. This is used in applications such as ambient light sensors in electronic devices.

   Humidity Sensors: The electrical properties of MOSFET can be influenced by humidity changes. Modified gate oxide can be employed to make MOSFETs responsive to variations in humidity levels, making them suitable for humidity sensing applications.

   Utilizing MOSFETs as sensors provides advantages such as high sensitivity, miniaturization potential, and compatibility with integrated circuit technology. These properties make them suitable for a wide range of sensing applications in diverse fields.

9. *Charges-Coupled Devices (CCDs):* MOSFETs are commonly used in the construction of CCDs, which are essential in imaging devices such as digital cameras and camcorders.

10. *Biomedical Devices:* FETs find applications in biomedical devices, such as biosensors and neural interfaces, where their electrical properties make them suitable for interfacing with biomedical systems.

These applications showcase the versatility of FETs in a wide range of electronic devices and systems across different industries.

6. Conclusion

This chapter offers a comprehensive review of transistors, encapsulating their evolutionary journey through four key perspectives: (1) the historical development of transistors, (2) constraints arising from scaling, (3) nanotransistors, and (4) contemporary and prospective applications shaping the present and future landscape of transistor utilization. In the realm of semiconductor technology, FET play a pivotal role, continuous evolving to meet the demands of modern electronics. This chapter explores the dynamic history of transistors, from the groundbreaking point contact transistor to the evolution of BJT, JFET, and the transformative MOSFET. The discussion encompasses limitations and solutions, including the introduction of FinFET and the promising potential of GAA-FET in overcoming scaling challenges.

While emphasizing the driving force of scaling in technological process, the chapter acknowledges the associated challenges like short channel effects and increased leakage current. Researchers are actively exploring alternative solutions, including alternating FET structures and materials. The exploring extends into the realm of nanotransistors, emphasizing their importance in the era of nanoelectronics and their critical role in sustain Moore's Law.

Lastly, the extensive applications of FET across diverse domains underscore their versatility in amplifiers, switching circuits, sensors and biomedical devices. The constant innovation and adaption in MOSFET technology are recognized as shaping the future of electronics, promising continued breakthroughs and advancements in the field.

Author details

Ankita Dixit
Independent Researcher, Pilani, Rajasthan, India

*Address all correspondence to: ankitadixit.199@gmail.com

References

[1] Shalf J. The future of computing beyond Moore's Law. Philosophical Transactions of the Royal Society A. 2020;**378**:20190061

[2] Thompson SE, Parthasarathy S. Moore's law: The future of Si microelectronics. Materials Today. 2006;**9**:20-25

[3] Hasegawa S, Grey F. Electronic transport at semiconductor surfaces—From point-contact transistor to micro-four-point probes. Surface Science. 2002;**500**:84-104

[4] Bondyopadhyay PK. The beginning junction transistor. Proceedings of the IEEE. 1998;**86**:63-77

[5] Ragavendran U, Ramachandran M. Low power and low area junction-less tunnel FET design. International Journal of Engineering & Technology. 2018;**7**:155-157

[6] Choi YC, Cha H-Y, Eastman LF, Michael G. Spencer: A new 4H-SiC normally off lateral channel vertical JFET with extremely low power losses: Source inserted double-gate structure with a supplementary highly doped region. IEEE Transactions on Electron Devices. 2005;**52**:1940-1948

[7] Rafin SMS, Hossain RA, Haque MA, Hossain MK, Haque MA, Mohammed OA. Power electronics revolutionized: A comprehensive analysis of emerging wide and ultrawide bandgap devices. Micromachines. 2023;**14**:2045

[8] Sahu A, Bramhane LK, Singh J. Symmetric lateral doping-free BJT: A novel design for mixed signal applications. IEEE Transactions on Electron Devices. 2016;**63**:2684-2690

[9] Nenadovic N, Nanver LK, Slotboom JW. Electrothermal limitations on the current density of high-frequency bipolar transistors. IEEE Transactions on Electron Devices. 2004;**51**:2175-2180

[10] Chen C, Labrousse D, Lefebvre S, Petit M, Buttay C, Morel H. Study of short-circuit robustness of SiC MOSFETs, analysis of the failure modes and comparison with BJTs. Microelectronics Reliability. 2015;**55**:1708-1713

[11] Levinzon FA. Noise of the JFET amplifier. IEEE Transactions on Circuits and Systems I: Fundamental Theory and Applications. 2000;**47**:981-985

[12] Baertsch RD, Engeler WE, Goldberg HS, Puckette CM. Fully integrated analog filters using bipolar-JFET technology. IEEE Journal of Solid-State Circuits. 1978;**13**:814-821

[13] Khanna VK. Short-channel Effects in MOSFETs: Integrated Nanoelectronics. New Delhi: Springer; 2016. pp. 73-93

[14] Kumar MJ. Controlling short-channel effects in deep-submicron SOI MOSFETs for improved reliability: A review. IEEE Transactions on Device and Materials Reliability. 2010;**4**:99-109

[15] D'Agostino F, Quercia D. Short-channel effects in MOSFETs. Introduction to VLSI Design (EECS 467). 2000;**70**:71-72

[16] Huq SI, Nafreen M, Rahman T, Bhadra S. Comparative study of full adder circuit with 32nm MOSFET, DG-FinFET and CNTFET. In: 4th International Conference on Advances in Electrical Engineering (ICAEE). Dhaka; 2017. pp. 38-43

[17] Wu M, Alivov YI, Morkoc H. High-κ dielectrics and advanced channel concepts for Si MOSFET. Journal of Materials Science: Materials in Electronics. 2008;**19**:915-951

[18] Long W, Ou H, Kuo JM, Chin KK. Dual-material gate (DMG) field effect transistor. IEEE Transactions on Electron Devices. 1999;**4**:865-870

[19] Hisamoto D et al. FinFET-a self-aligned double-gate MOSFET scalable to 20 nm. IEEE Transactions on Electron Devices. 2000;**47**:2320-2325

[20] Dixit A, Gupta N. A compact model of gate capacitance in ballistic gate-all-around carbon nanotube field effect transistors. International Journal of Engineering. 2021;**34**:1718-1724

[21] Das UK, Bhattacharyya TK. Opportunities in device scaling for 3-nm node and beyond: FinFET versus GAA-FET versus UFET. IEEE Transactions on Electron Devices. 2020;**67**:2633-2638

[22] Cutress I. Intel's Manufacturing Roadmap from 2019 to 2029: Back Porting, 7 nm, 5 nm, 3 nm, 2 nm, and 1.4 nm. AnandTech; 11 Dec 2019

[23] Das UK, Bhattacharyya TK. Opportunities in device scaling for 3-nm node and beyond: FinFET versus GAA-FET versus UFET. IEEE Transactions on Electron Devices. 2020;**67**:2633-2638

[24] Devi D, Rakesh P, Rakesh V. Impact of scaling gate insulator thickness on the performance of carbon nanotube field effect transistors (CNTFETs). Journal of Nano-and Electronic Physics. 2013;**5**:02014-02019

[25] Yu L, Chang S, Ahmed H, Wang S, Bell CY, Yang C, et al. FinFET scaling to 10 nm gate length. In: Digest International Electron Devices Meeting. San Francisco, USA; 2002. pp. 251-254

[26] Gupta N, Dixit A. Carbon Nanotube Field-Effect Transistors (CNFETs): Structure, Fabrication, Modeling, and Performance. Carbon Nanomaterial Electronics: Devices and Applications. Singapore: Springer Singapore; 2021. pp. 199-214

[27] Lundstrom MS. Fundamentals of Nanotransistors. Vol. 6. Indiana, USA: World Scientific Publishing Company; 2016

[28] Li J, Li Y, Zhou N, Xiong W, Wang G, Zhang Q, et al. Study of silicon nitride inner spacer formation in process of gate-all-around nano-transistors. Nanomaterials. 2020;**10**:793

[29] Timp G, Bude J, Bourdelle KK, Garno J, Ghetti A, Gossmann H, et al. The ballistic nano-transistor. In: International Electron Devices Meeting. USA; 1999. pp. 55-58

[30] Singh S et al. Implementation of gate-all-around gate-engineered charge plasma nanowire FET-based common source amplifier. Micromachines. 2023;**14**:1357

[31] Yim S-M, Kk O. Switched resonators and their applications in a dual-band monolithic CMOS LC-tuned VCO. IEEE Transactions on Microwave Theory and Techniques. 2006;**54**:74-81

[32] Nouri R, Guan W. Nanofluidic charged-coupled devices for controlled DNA transport and separation. Nanotechnology. 2021;**32**:345501

[33] Sreenivasulu V, Narendar V. Design and temperature assessment of junctionless nanosheet FET for nanoscale applications. Silicon. 2022;**14**:3823-3834

[34] Mishra AK, Jarwal DK, Mukherjee B, Kumar A, Ratan S, Jit S. CuO nanowire-based extended-gate

field-effect-transistor (FET) for pH sensing and enzyme-free/receptor-free glucose sensing applications. IEEE Sensors Journal. 2020;**20**:5039-5047

[35] Dixit A, Gupta N, Chaturvedi N. Ab initio study of carbon nanotube field effect transistor gas sensor for detection of Ammonia and nitrogen dioxide gas. IEEE Transactions on Nanotechnology. 2022;**21**:564-574

[36] Tiwari S, Saha R. Improved optical performance in near visible light detection photosensor based on TFET. Microelectronics Journal. 2022;**129**:105554

[37] Sedki M, Shen Y, Mulchandani A. Nano-FET-enabled biosensors: Materials perspective and recent advances in North America. Biosensors and Bioelectronics. 2021;**176**:112941

[38] Li H, Li D, Chen H, Yue X, Fan K, Dong L, et al. Application of silicon nanowire field effect transistor (SiNW-FET) biosensor with high sensitivity. Sensors. 2023;**23**:6808

[39] Mishra AK, Jarwal DK, Mukherjee B, Kumar A, Ratan S, Jit S. CuO nanowire-based extended-gate field-effect-transistor (FET) for pH sensing and enzyme-free/receptor-free glucose sensing applications. IEEE Sensors Journal. 2020;**20**:5039-5047

[40] Dai C, Liu Y, Wei D. Two-dimensional field-effect transistor sensors: The road toward commercialization. Chemical Reviews. 2022;**122**:10319-10392

[41] Gelao G, Marani R, Perri AG. Effect of CNT parameters variations on CNTFET amplifier performance. ECS Journal of Solid State Science and Technology. 2023;**12**:011004

Chapter 2

A Compact Model of DMG-GC-DOTTDCD Cylindrical Gate MOSFET

Hind Jaafar, Abdellah Aouaj, Ahmed Bouziane and Benjamin Iñiguez

Abstract

A comprehensive analysis of the surface potential, threshold voltage, and subthreshold swing (SS) for dual-metal gate graded channel and dual oxide thickness with two different dielectric constants cylindrical gate (DMG-GC-DOTTDCD) metal-oxide-semiconductor field-effect transistors (MOSFETs) is proposed to investigate short-channel effects (SCEs). An analytical model for drain current is developed employing a quasi-two-dimensional cylindrical form of the Poisson equation and is expressed as a function of the surface potential, which is determined using the current density expressions. The DMG-GC-DOTTDCD device is compared with DMG-GC-DOT, revealing that our structure provides superior immunity against short-channel effects (incorporated in the drain current model). The analytically obtained results align with those achieved through the Silvaco Atlas-TCAD software.

Keywords: dual-metal gate, dual oxide thickness (DOT), graded channel, drain current, subthreshold current

1. Introduction

The reduction in the dimensions of MOSFET transistors is not arbitrary but follows a systematic scaling law [1]. This law, an improved version of Dennard's initial 1974 law, aims to quantitatively describe key parameters of technology (such as dimensions, doping, capacitance, and current) using a single factor, denoted as K. The purpose is to facilitate the prediction of expected performance for future technological nodes [2].

However, the downsizing of transistors also gives rise to an increase in parasitic effects. Notable among these effects are short-channel effects (resulting in a decrease in the transistor's threshold voltage, DIBL, etc.) [3], gate leakage current, and technological fluctuations (variations in doping, thickness, etc.). These effects significantly disrupt the operation of integrated circuits.

Hence, there is a growing need to explore new component architectures and alternative materials beyond the conventional ones used in microelectronics (such as Si, SiO_2, and polycrystalline silicon). This exploration aims to introduce innovations while minimizing deviations from existing well-established manufacturing processes. Various device types are currently under investigation in applied and fundamental research, attracting attention from major integrated circuit manufacturers. Examples include silicon-on-insulator (SOI) devices and multi-gate transistors, such as double gate (DG), all around (GAA), and surrounding gate (SG). These emerging architectures should offer improved control over the channel potential through gate voltage manipulation, thereby extending the limits of MOSFET miniaturization. Notably, the gate-all-around MOSFET, also known as the "surrounding-gate MOSFET" [4], demonstrates enhanced electrostatic potential control compared to the DG MOSFET structure [5].

In recent years, numerous studies have been conducted on surrounding gate (SG) MOSFETs to mitigate short-channel effects (SCEs) and enhance hot carrier reliability. Several investigations propose gate material engineering as a viable solution to address these challenges. The utilization of a dual-material gate (DMG) structure, employing two metals with different work functions, has been shown to improve SCEs compared to the single material (SM) SG MOSFET [6].

Channel engineering, particularly graded channel (GC) structures [7], has been identified as another effective approach to suppress SCEs and elevate device performance. Incorporating a GC involves creating two doping regions—an extensively doped region near the source end and a lightly doped region near the drain end. This configuration has demonstrated a significant enhancement in hot carrier reliability and resistance against SCEs.

Additionally, scientists have investigated the utilization of high-k dielectrics as a substitute for SiO2 in the role of gate dielectric. This substitution is intended to minimize gate leakage current and improve the control of the gate over the channel [8, 9].

We introduce a concise drain-current model designed for the short-channel DMG-GC-DOTTDCD SRG MOSFET illustrated in **Figure 1** and this structure compared to DMG–GC–DOT [10]. The model is constructed upon analytical expressions of surface potentials derived from the long-channel SRG MOSFET model. Specifically, the drain-current model is formulated to capture short-channel effects (SCEs), including drain-induced barrier lowering (DIBL), VT roll-off, and subthreshold slope degradation. These SCEs are represented through straightforward analytical expressions [11].

The surface potentials are computed using a simplified parabolic function, showcasing its practicality in various engineering applications. Our model intentionally excludes quantum effects and mobility degradation. Quantum effects become pronounced when the radius of the cylindrical body is less than 5 nm [12]. To verify the accuracy of our model, the calculated potentials and drain currents are compared to 3D simulations obtained through ATLAS-TCAD.

2. Development of analytical model

2.1 Model derivation

Figure 1b displays a cross-sectional view along the channel direction of the DMG-GC-DOTTDCD MOSFET. Conceptually, a dual-material gate device can be visualized as two sub-devices connected in series. Two metal gates denoted as M_1 and M_2, with distinct work functions, are present, where the work function of M_1 is higher than M_2 ($\varphi_1 \succ \varphi_2$). In the halo region ($L_1$) of the device, the doping concentration (N_H) is

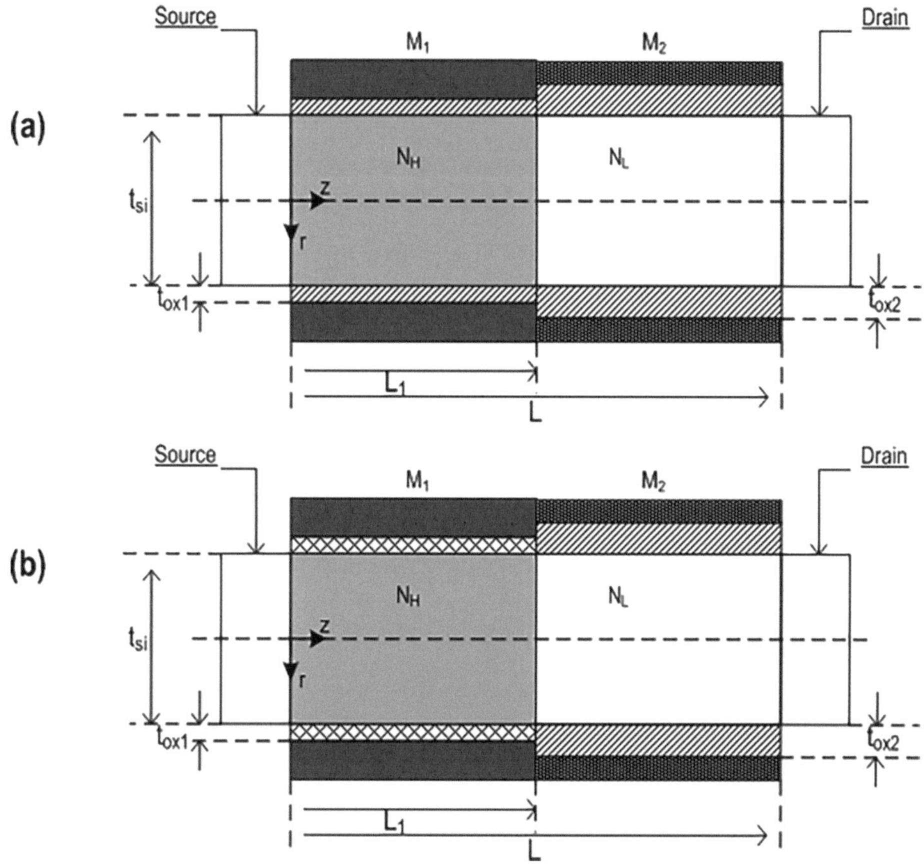

Figure 1.
Cross-sectional perspectives illustrating diverse device design configurations in SG MOSFET. (a) (DMG–GC–DOT), (b) (DMG–GC–DOTTDCD).

higher compared to the rest of the channel ($L_2 = L - L_1$). Additionally, the oxide thickness t_{ox2} (SiO2) in the remaining channel region ($L_2 = L - L_1$) is greater than t_{ox1} (high-k) the thickness of the oxide (high-k) in region (L_1) [12].

The cylindrical symmetry of the device necessitates the adoption of a cylindrical coordinate system, which includes a radial direction (r) and a horizontal direction (z), omitting the angular component in the figure. The structural symmetry guarantees that both the potential and electric field show no angular variation within the plane of the radial direction. As a result, a 2D analysis is deemed sufficient.

2.2 Surface potential model

Determining the distribution of electrostatic potential and electric field within the silicon channel necessitates solving Poisson's equation. By excluding the influence of charge carriers and fixed charges, Poisson's equation can be represented in cylindrical coordinates for two separate regions (i = 1, 2):

$$\frac{1}{r}\frac{\partial}{\partial}\left(r\frac{\partial\varphi_i(r,z)}{\partial r}\right)+\frac{\partial^2\varphi_i(r,z)}{\partial z^2}=\frac{qN_i}{\varepsilon_{si}} \qquad (1)$$

$$0 \le z \le L, 0 \le r \le \frac{t_{si}}{2}$$

The thickness of the silicon channel is denoted as t_{si}, where ε_{si} represents the dielectric constant of the silicon pillar, and $N_1 = N_H$ while $N_2 = N_L$.

The assumption is made that the potential distribution in the two regions follows a parabolic profile [13] in the radial direction, expressed as:

$$\varphi_i(r, z) = p_{i0}(z) + p_{i1}(z).r + p_{i2}(z).r^2 \quad (2)$$

The functions $p_{i0}(z)$, $p_{i1}(z)$ and $p_{i2}(z)$solely depend on the variable z.

Because of its symmetry, the electric field at the center of the silicon pillar is null.

$$\left(\frac{\partial \phi_i(r, z)}{\partial r}\right)_{r=0} = 0 = p_{i1}(z)$$

The interface between the oxide and silicon upholds the continuity of electric flux.

$$\left(\frac{\partial \phi_i(r, z)}{\partial r}\right)_{r=\frac{t_{si}}{2}} = \frac{c_{ox_i}}{\varepsilon_{si}}(V_{GS} + V_{FB_i} + \phi_{si}(z)) = p_{i2}(z) t_{si}$$

where $c_{oxi} = \frac{2\varepsilon_{oxi}}{t_{si} \ln\left(1 + \frac{2t_{oxi}}{t_{si}}\right)}$

The oxide capacitance for each oxide section (i = 1, 2) is denoted as Coxi. represents the gate-to-source voltage. φi(r, z) represents the surface potential, εox1 and εox2 are the dielectric constants for the high-k and SiO2 gate oxides, respectively. In addition, tox1 signifies the thickness of the oxide layer in region L1, and tox2 pertains to the thickness of the oxide layer in the region L-L1.

The flat band voltages for the two regions, identified as V_{FBi}, will display disparities, and their specific values are outlined below: $V_{FB_1} = \phi_1 - \phi_{siH}$,

$$V_{FB_2} = \phi_2 - \phi_{siL}.$$

Here, φ_1 and φ_2 represent the work functions of M_1 and M_2, respectively. Additionally, φ_{siH} and φ_{siL} correspond to the work functions of region L_1 and the rest of the silicon pillar, respectively.

Through the resolution of the Poisson equation in the two regions and the inclusion of boundary conditions, it is streamlined into the subsequent expression:

$$\frac{\partial^2 \phi_{si}(z)}{\partial z^2} - \lambda_i^2 \phi_{si}(z) = D_i \text{ i} = 1, 2 \quad (3)$$

where $\lambda_i = \sqrt{4c_{oxi}/\varepsilon_{si} t_{si}}$, is the characteristic length and $D_i = \frac{qN_i}{\varepsilon_{si}} - \lambda_i^2(V_{GS} - V_{FBi})$

The potential at the source end is $\varphi_1(0, 0) = V_{bi1}$, where V_{bi1} is the built-in potential.

The potential at the drain end is $\varphi_2(0, L) = V_{bi2} + V_{DS}$, where L is the device channel length and V_{DS} is the drain to source voltage.

The surface potential's general solution is represented in the following form:

$$\phi_{si}(z) = A_i \exp(-\lambda_i z) + B_i \exp(\lambda_i z) - \frac{D_i}{\lambda_i^2} \quad (4)$$

By applying boundary conditions, it is possible to ascertain the coefficients A_i and B_i (for i = 1, 2):

$$A_1 = V_{bi_1} + \frac{D_1}{\lambda_1{}^2} - B_1$$

$$A_2 = \left(\left(V_{bi_2} + V_{ds} + \frac{D_2}{\lambda_2{}^2}\right) - B_2 \exp(\lambda_2 L)\right) \exp(\lambda_2 L)$$

$$B_2 = \frac{U_2}{U_0},\ B_1 = \frac{U_1}{U_0},\ U_2 = a_0 C_2 - c_0 C_1$$

$$U_1 = d_0 C_1 - b_0 C_2,\ U_0 = a_0 d_0 - b_0 c_0,$$

$$C_1 = \left(V_{bi_2} + V_{ds} + \frac{D_2}{\lambda_2{}^2}\right)\exp(-\lambda_2 L_1) - \left(V_{bi_1} + \frac{D_1}{\lambda_1{}^2}\right)\exp(-\lambda_2 L)\exp(-\lambda_1 L_1) + \left(\frac{D_1}{\lambda_1{}^2} - \frac{D_2}{\lambda_2{}^2}\right)\exp(-\lambda_2 L)$$

$$C_2 = -\lambda_2\left(V_{bi_2} + V_{ds} + \frac{D_2}{\lambda_2{}^2}\right)\exp(-\lambda_2 L_1) + \lambda_1\left(V_{bi_1} + \frac{D_1}{\lambda_1{}^2}\right)\exp(-\lambda_2 L)\exp(-\lambda_1 L_1)$$

Figure 2 illustrates the variation of surface potential along the channel for both DMG-GC-DOT and DMG-GC-DOTTDCD. Notably, the inclusion of two oxide thicknesses with distinct dielectric constants in the (DMG-GC-DOTTDCD) SG MOSFET leads to an elevation in the potential barrier. Additionally, the figure unmistakably indicates that the lowest surface potential occurs in the initial region near the source of the DMG-GC-DOTTDCD.

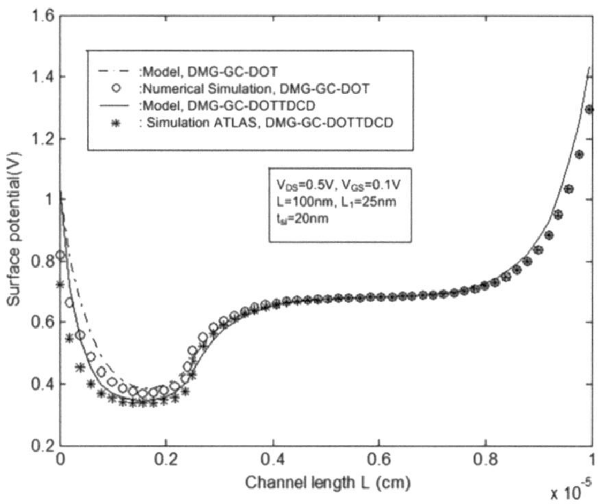

Figure 2.
Surface potential variation along the channel for DMG-GC-DOT and DMG-GC-DOTTDCD configurations.

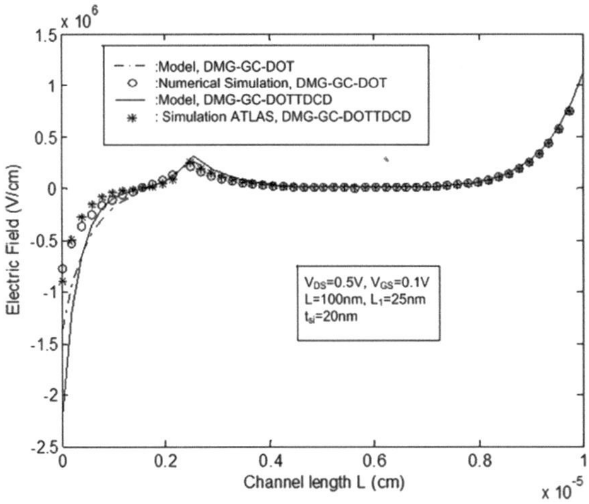

Figure 3.
Variation of the electric field versus channel length for DMG-GC-DOT and DMG-GC-DOTTDCD with V_{GS} = 0.1 V and V_{DS} = 0.5 V.

By computing the derivative of the surface potential $\varphi_{si}(r = R, z)$ with respect to z, we can express the electric field E (z) at the channel surface in the z-direction as:

$$E_i(z) = -A_i\lambda_i \exp(-\lambda_i z) + B_i\lambda_i \exp(\lambda_i z) \quad (5)$$
$$0 \leq z \leq L, \mathrm{i} = 1, 2$$

In **Figure 3**, a distinct step change in the potential is evident. This profile showcases a noticeable alteration in the electric field precisely situated at the junction of the two metals. The heightened lateral electric field observed in the channel beneath the interface of the two gate materials contributes to increased efficiency in carrier transport.

2.3 Threshold voltage model

In the structure of a DMG-GC-DOTTDCD MOSFET, the minimum surface potential consistently resides beneath the gate material with the higher work function (M_1). Consequently, determining the position of the minimum surface potential involves setting the derivative of the surface potential under M_1 to zero [14]. By equating $\frac{d\varphi_{s1}}{dz}$ to zero, the following expression is obtained:

$$z_{\min} = \frac{1}{2\lambda}\sqrt{\frac{B_1}{A_1}}$$

The threshold voltage V_{TH} is characterized by the point where the minimum surface potential, $\varphi_{si,\min}$, reaches a value of $2\varphi_B$, with V_{GS}being the gate-source voltage and φ_B representing the bulk Fermi potential.

We investigated the minimum surface potential within the highly doped region (L_1), characterized by significant N_H concentration (region 1).

The value of $\varphi_{si,\min}$ can be inferred from Eq. (4):

$$\varphi_{si,\,\min} = 2\sqrt{A_1 B_1} - \frac{D_1}{{\lambda_1}^2} \tag{6}$$

The threshold voltage can be expressed as:

$$V_{TH} = \left(-\eta + \sqrt{\eta^2 - 4\sigma\xi}\right)/2\sigma$$

where:

$$a_0 = 2\exp(-\lambda_2 L)\sinh(\lambda_1 L_1)$$
$$b_0 = 2\sinh(\lambda_2(L - L_1))$$
$$c_0 = 2\lambda_1\exp(-\lambda_2 L)\cosh(\lambda_1 L_1)$$
$$d_0 = -2\lambda_2\cosh(\lambda_2(L - L_1)),$$
$$U_0 = a_0 d_0 - b_0 c_0 e_0 = \exp(-\lambda_1 L_1)e_1 = \exp(-\lambda_2 L_1)e_2 = \exp(-\lambda_2 L)$$
$$a_1 = \frac{(qN_{aH})}{\varepsilon_{si}{\lambda_1}^2} + V_{FB_1}, a_2 = \frac{(qN_{aL})}{\varepsilon_{si}{\lambda_2}^2} + V_{FB_2}, b_1 = e_2 e_0(V_{bi_1} + a_1)$$
$$b_2 = e_1(V_{ds} + V_{bi_2} + a_2,)c_1 = \lambda_1 b_1, c_2 = \lambda_2 b_2$$
$$E = (d(e_2 e_0 - e_1) + b(\lambda_1 e_2 e_0 - \lambda_2 e_1))/U_0$$
$$\sigma = -(4E^2 + 4E + 1), \eta = 4(V_{bi_1} + a_1) + E4D - 8DE + 2a_1 + 4\phi_B$$
$$\xi = 4(V_{bi_1} + a_1)D - 4D^2 - {a_1}^2 - 4a_1\phi_B - 4{\phi_B}^2$$

In **Figure 4**, we illustrate the variation in threshold voltage shift (ΔV_{TH}) with respect to channel length for both DMG-GC-DOT and DMG-GC-DOTTDCD MOSFETs. It is clear from the figure that DMG-GC-DOTTDCD MOSFET exhibits greater effectiveness in (ΔV_{TH}) compared to DMG-GC-DOT MOSFETs.

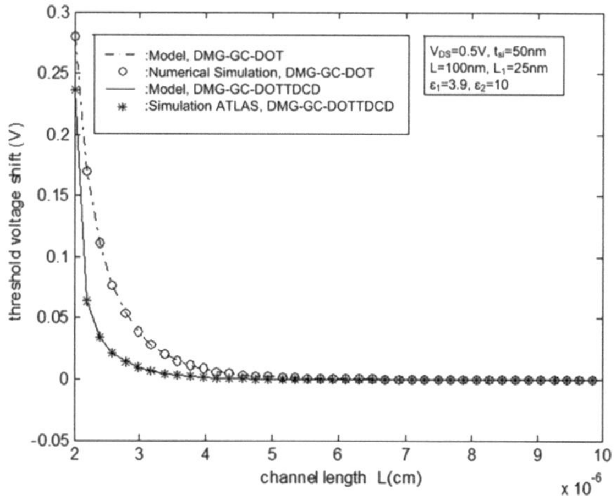

Figure 4.
Threshold voltage shift versus channel length for DMG-GC-DOT and DMG-GC-DOTTDCD MOSFET.

2.4 Subthreshold swing

We examined the minimum surface potential in the high doping concentration region (L_1, region 1) [9].

An essential parameter, the subthreshold swing (SS), is defined as:

$$SS = \frac{KT}{q}\ln(10)\left(\frac{\partial\phi_{si,\min}}{\partial V_{GS}}\right)^{-1} \tag{7}$$

From (7), we obtain:

$$\left(\frac{\partial\varphi_{si,\min}}{\partial V_{GS}}\right) = 1 + (A_1B_1)^{\left(\frac{-1}{2}\right)}(A_1P_1 + B_1P_2)$$

where

$$\begin{aligned} P_1 = {} & \frac{d_0}{U_0}(\exp(-(\lambda_2 L + \lambda_1 L_1)) - \exp(\lambda_2 L_1)) \\ & - \frac{b_0}{U_0}\lambda_2(\exp(-\lambda_2 L_1) - \exp(-(\lambda_2 L + \lambda_1 L_1))) \end{aligned}$$

$$P_2 = -1 - P_1$$

Figure 5 illustrates the variation in subthreshold swing across the channel for both DMG-GC-DOT and DMG-GC-DOTTDCD. It is apparent that the subthreshold swing is lower for the DMG-GC-DOTTDCD device in comparison to the DMG-GC-DOT.

Figure 6 illustrates the variations in drain-induced barrier lowering (DIBL) for both DMG-GC-DOT MOSFET and DMG-GC-DOTTDCD MOSFET concerning the channel length. DIBL is quantified by $\Delta V_{th}/\Delta V_{ds}$, where $\Delta V_{th} = V_{th}|_{(V_{ds}=0)} - V_{th}|_{(V_{ds}=2)}$ and $\Delta V_{DS} = 2V$. The graph indicates that due to the combined influence of dual oxide thicknesses with distinct dielectric constants, DMG-GC-DOTTDCD exhibits superior suppression of DIBL compared to DMG-GC-DOT.

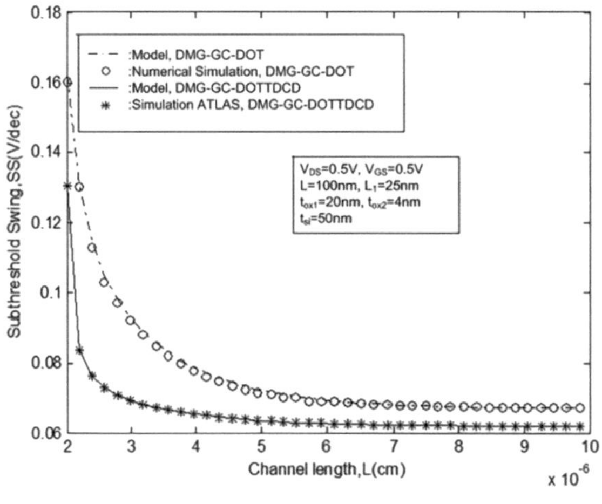

Figure 5.
Subthreshold swing (SS) versus channel length (L) for DMG-GC-DOT and DMG-GC-DOTTDCD.

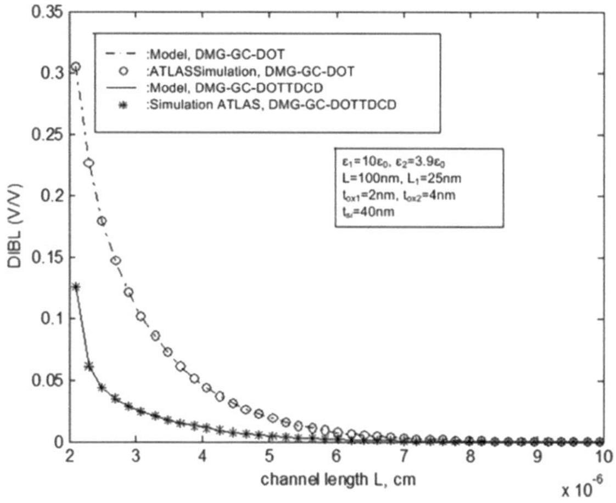

Figure 6.
DIBL variations versus the channel length for DMG-GC-DOT MOSFET and DMG-GC-DOTTDCD MOSFET.

2.5 Current model

2.5.1 Linear region

To formulate a succinct drain current model based on charge density, the method entails calculating separate drain currents in the regions near the source and drain. Subsequently, an equation for the drain current in the linear region is derived. In the strong inversion region, where the drain current is primarily affected by the drift, it can be expressed as follows [15, 16]:

$$I_{ds}(z) = \pi t_{si} Q_{ni}(z) \frac{\mu_{ni}(dV(z)/dz)}{1 + (1/E_{sat})(dV(z)/dz)} \tag{8}$$

Here, t_{si} represents the device radius, $dV(z)/dz$ denotes the channel electric field along the z-direction, V(z) signifies the channel potential along the z-direction, E_{sat} is the critical field given by $2\upsilon_{sat}/\mu_{ni}$, where υ_{sat} represents the saturation velocity, and $Q_{ni}(z)$ stands for the surface charge density at a specific point z in the strong inversion region:

$$Q_{ni}(z) = c_{ox_i}\left(\mathrm{V}_{gs} - V_{thi} - V(z)\right) \tag{9}$$

The gate oxide capacitance in the DMG-GC-DOTTDCD MOSFET is represented as c_{ox_i} for both regions.

Here, V_{thi} represents the threshold voltage for the respective regions with i equal to 1 and 2. Additionally, V_{gs} represents the gate-to-source bias.

μ_{ni} signifies the mobility in region i (i = 1, 2) [17], and it is expressed as:

$$\mu_{ni} = \frac{\mu_1}{1 + \theta_i\left(\mathrm{V}_{gs} - V_{thi}\right)} \tag{10}$$

where μ_1 is doping-dependent mobility given a

$$\mu_1 = \frac{\mu_{n0}}{\sqrt{\left(1 + \left[N_i/\left(N_{ref} + (N_i/S)\right)\right]\right)}} \tag{11}$$

Here, μ_{n0} represents the low-field mobility and θ_i is a fitting parameter. The parameters S and N_{ref} represent the balance between phonon and impurity scattering, encapsulating their respective trade-offs, where S is set at 450 and N_{ref} is equal to $3 \times 10^{22} m^{-3}$.

By substituting Eqs. (9)–(11) into Eq. (8) and performing the integration, the resulting expression for the drain current in the linear region is as follows:

$$I_{ds} = \frac{\pi t_{si} \mu_{eff} c_{ox_1} \left[V_{gs} - V_{th} - (V_{ds}/2)\right] V_{ds}}{L\left(1 + (V_{ds}/LE_{eff})\right)} \tag{12}$$

where

$$\mu_{eff} = \frac{L}{[(L_1/\mu_{n1}) + (L_2/\mu_{n2})]}$$

$$E_{eff} = \frac{2v_{sat}}{\mu_{eff}}$$

2.5.2 Saturation region

The saturation region drain current (I_{dsSat}) is expressed as [14]:

$$I_{dsSat} = \pi t_{si} v_{sat} Q_{inSat} \tag{13}$$

where Q_{inSat} inversion charge at $V_{ds} = V_{dsSat}$ and given as

$$Q_{inSat} = c_{ox_i}\left(V_{gs} - V_{th} - V_{dsSat}\right) \tag{14}$$

where V_{dsSat} denotes the drain saturation voltage.

Upon substituting Eq. (14) into Eq. (13), the expression for the drain current in the saturation region can be written as:

$$I_{dsSat} = \pi t_{si} c_{ox_i} v_{sat}\left(V_{gs} - V_{th} - V_{dsSat}\right) \tag{15}$$

The value of V_{dsSat} can be obtained by setting V_{ds} to V_{dsSat} and equating Eqs. (12) and (15), yielding:

$$V_{dsSat} = \frac{V_{gs} - V_{th}}{1 + \left(V_{gs} - V_{th}\right)/LE_{eff}} \tag{16}$$

where V_{th} is the threshold voltage of DMG-GC-DOTTDCD MOSFET [6].

2.5.3 Subthreshold current

The electron current along the channel of a surrounding gate MOSFET can be expressed as detailed in:

$$I_{sub}(z) = \pi t_{si} \int_0^{t_{si}/2} J(r,z)dr \tag{17}$$

The dominant factor in subthreshold conduction is the diffusion current, expressed as:

$$J(r,z) = -q \frac{\mu'_n}{1+\theta(V_{gs} - V_{th})} n(r,z) \frac{dV(z)}{dz} \tag{18}$$

where

$$n = n_i e^{\varphi_{s,\min} - V/V_T}$$

Using this derived value for the inversion charge carriers [18]

$$I_{sub}(z) = \frac{\mu'_n}{1+\theta(V_{gs} - V_{th})} \pi q n_i \left(\frac{t_{si}}{2}\right)^2 \frac{dV(z)}{dz} e^{\frac{\varphi'_{s,\min} - V}{V_T}} \tag{19}$$

Through the integration of the given equation across the channel and accounting for the boundary conditions at the source $V(0) = 0$ and drain $V(L) = V_{ds}$, where the potential in region 2 exceeds that in region 1, we deduce:

$$I_{sub} = \frac{\mu'_{n1}}{1+\theta(V_{gs} - V_{th1})} \frac{\pi t_{si}{}^2 n_i{}^2 kT \left(1 - e^{\frac{-qV_{ds}}{kT}}\right) e^{\frac{q\varphi_{\min,1}}{kT}}}{N_1 L_1} \tag{20}$$

And the doping-dependent mobility (μ'_{ni}) is given as:

$$\mu'_{ni} = \frac{\mu_n}{\sqrt{\left(1 + \frac{N_i}{(N_{ref} + N_i S_1}\right)}} \tag{21}$$

Fitting parameters S_1 and θ, where μ_n represents electron mobility, are crucial in achieving a satisfactory match between modeled and simulated results in the analysis. To achieve this alignment, the values of the fitting parameters were set to 450 for S_1 and 0.04 for θ.

2.5.4 A complete model for all regimes

To formulate a unified expression covering all operational regimes, the introduction of a transition function is necessary [18, 19]. A hyperbolic tangent function is well-suited for this purpose, allowing us to ultimately represent the overall drain current, denoted as I_{tot} model, as follows:

$$I_{tot} = I_{ds} \times \frac{1}{2}[1 - \tanh[G(V_{gs} - V_{th1})]] + I_{sub} \times \frac{1}{2}[1 + \tanh[G(V_{gs} - V_{th1})]]$$

A new parameter, denoted as the transition parameter G, is introduced in the equation. The threshold voltage V_{th1} delineates the midpoint of transition between two hyperbolic tangent functions. The parameter G regulates the sharpness of the transition around V_{th1}, with a predefined value of 0.5.

Figure 7 illustrates the comparison between the modeled expression and TCAD simulation of the subthreshold drain current (I_{DS}) variation concerning V_{gs} for DMG-GC-DOT and DMG-GC-DOTTDCD MOSFETs. It is observed that DMG-GC-DOTTDCD yields a higher I_{DS} compared to DMG-GC-DOT. However, it is noteworthy that an elevation in the subthreshold drain current leads to an increase in the subthreshold leakage current and a reduction in the subthreshold swing, factors that require minimization for applications involving ultra-low power devices.

2.6 Transconductance formulation

The transconductance is a vital parameter in analog circuit design, calculated by taking the derivative of the drain current with respect to V_{gs}. This calculation enables the verification of the model's differential behavior and is formulated as:

$$g_m = \left(\frac{dI_{ds}}{dV_{gs}}\right)\bigg|_{V_{ds}=const} \tag{22}$$

By utilizing the drain current equations along with Eq. (22), the expression for transconductance (g_m) in the linear region is derived as presented below:

$$g_m = \pi t_{si} c_{oxi} V_{ds}\left[\frac{Ab - aB}{\left(L\left(1+\left(\frac{V_{ds}}{LE_{eff}}\right)\right)\right)^2}\right] \tag{23}$$

where

$$a = \mu_{eff}\left(V_{gs} - V_{th} - \frac{V_{ds}}{2}\right)$$

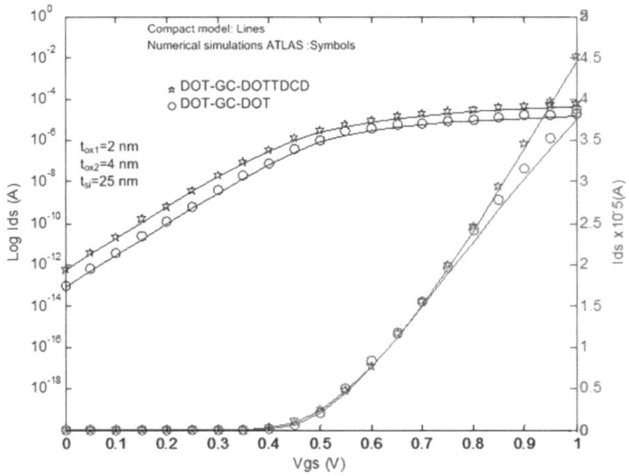

Figure 7.
The drain current of the DMG-GC-DOT MOSFET is plotted against gate voltage and compared with DMG and DMG-DOT counterparts, presented in both linear and logarithmic scales. Compact model: lines; numerical simulations ATLAS: Symbols.

$$b = L\left(1+\left(\frac{V_{ds}}{LE_{eff}}\right)\right), E = \frac{-L\theta_i{\mu_1}^3(L_1+L_2)}{\left[(L_1+L_2)(\mu_1+\theta_i\mu_1 V_{gs}) - \theta_i\mu_1(L_1V_{th1}+L_2V_{th2})\right]^2}$$

$$A = E\left(V_{gs} - V_{th} - \frac{V_{ds}}{2}\right) + \mu_{eff}, B = \frac{V_{ds}}{2v_{sat}}E\mu_{ni} = \frac{\mu_1}{1+\theta_i(V_{gs} - V_{thi})}$$

In this context, θ_i serves as a parameter for fitting, while μ_1 represents the empirically derived relationship expressing the dependency of mobility on impurity density:

$$\mu_1 = \frac{\mu_{n0}}{\sqrt{(1+\left[N_i/(N_{ref}+(N_i/S))\right])}}$$

$$\mu_{eff} = \frac{L}{[(L_1/\mu_{n1}) + (L_2/\mu_{n2})]}, \quad E_{eff} = \frac{2v_{sat}}{\mu_{eff}}$$

where μ_{n0} is the electron mobility. The parameters S and N_{ref} reflect the balance between phonon and impurity scattering, with values specified as S = 450 and $N_{ref} = 3 \times 10^{22} m^{-3}$.

Figure 8 depicts a comparison of transconductance versus gate-source voltage characteristics, calculated using both the DMG-GC-DOT and DMG-GC-DOTTDCD models incorporating ε_1 and ε_2. These results are then contrasted with simulated data, indicating a significant degree of similarity between them.

2.7 Drain conductance formulation

The conductance is a crucial parameter in simulating analog circuits and is defined as follows:

$$g_{ds} = \left(\frac{dI_{ds}}{dV_{ds}}\right)\Bigg|_{V_{gs}=const} \tag{24}$$

By employing the drain current equation, the expression for output conductance (g_{ds}) is derived as follows:

$$g_{ds} = \frac{Cd - Dc}{\left(L\left(1+\frac{V_{ds}}{LE_{eff}}\right)\right)^2} \tag{25}$$

where

$$C = \pi t_{si}\mu_{eff}c_{oxi}(V_{gs} - V_{th} - V_{ds}), D = \frac{1}{E_{eff}}$$

$$c = \pi t_{si}\mu_{eff}c_{oxi}V_{ds}\left(V_{gs} - V_{th} - \frac{V_{ds}}{2}\right), d = L\left(1+\frac{V_{ds}}{LE_{eff}}\right)$$

The graph in **Figure 9** illustrates the change in drain conductance concerning the drain-source voltage for both DMG-GC-DOT and DMG-GC-DOTTDCD structures,

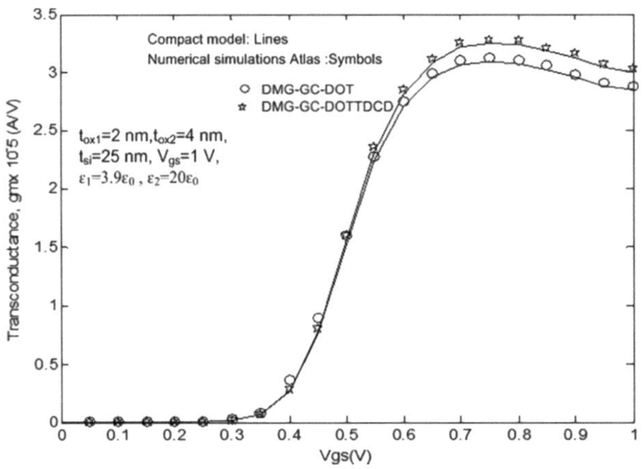

Figure 8.
Comparison between DMG-GC-DOT and ε_1 and ε_2 in DMG-GC-DOTTDCD structure for transconductance.

with ε_1 and ε_2 representing parameters in the latter. An analysis reveals that the drain conductance in the DMG-GC-DOTTDCD structure surpasses that in the DMG-GC-DOT structure.

In **Figure 10**, the transconductance (g_m) exhibits a consistent variation in response to the gate-source voltage (V_{gs}), considering both low and high drain voltages ($V_{ds} = 0.05\ V$ and 1V). A comparison with numerical simulations reveals good agreement for both low and high drain voltage conditions.

Figure 11 illustrates the drain conductance (g_{ds}) in relation to the drain voltage (V_{ds}) for different gate voltage values ($V_{gs} = 0.5\ V$, 1V, and 1.5V). The model's accuracy is affirmed by the close alignment observed between analytical and simulated outcomes.

In **Figure 12**, the capacitance characteristics, namely C_{sg} and C_{dg}, are showcased as functions of the gate voltage, spanning various drain-source voltage values. The modeled capacitance displays a smooth transition across different operational regimes.

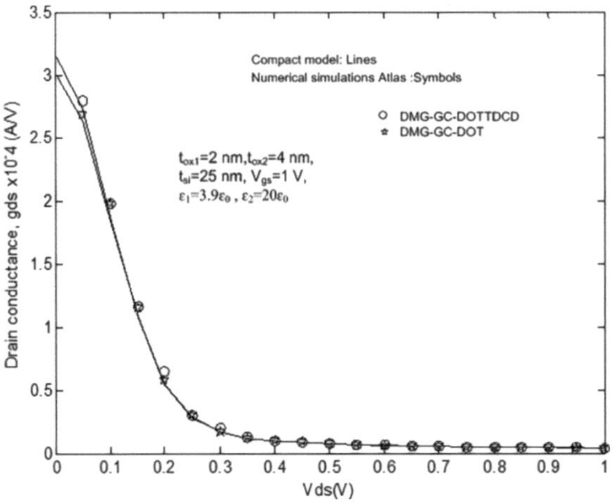

Figure 9.
Comparison between DMG-GC-DOT and ε_1 and ε_2 in DMG-GC-DOTTDCD structure for drain conductance.

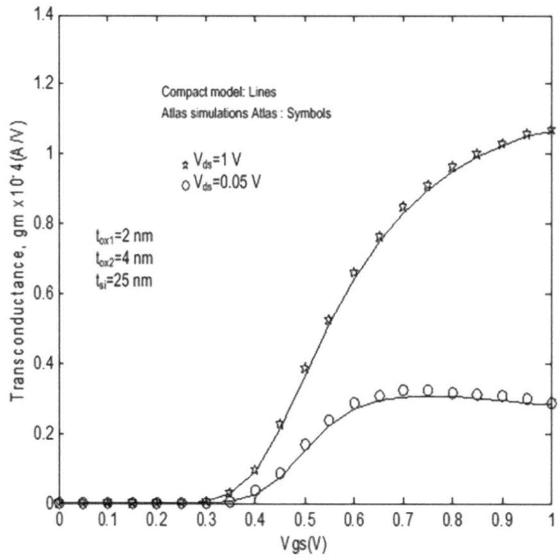

Figure 10.
Transconductance versus gate-source voltage for different value of V_{ds} in DMG-GC-DOTTDCD MOSFET structure.

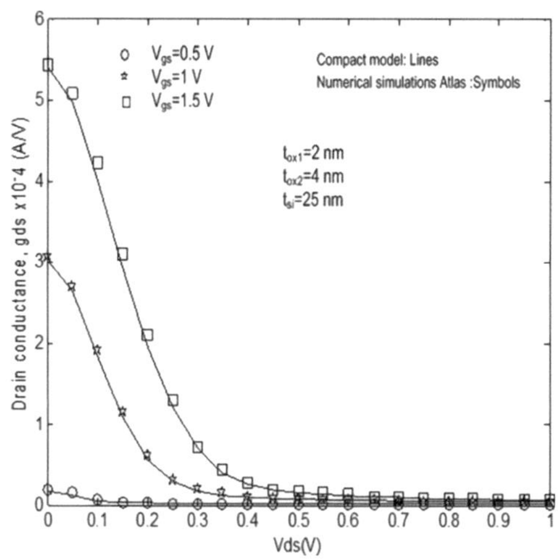

Figure 11.
Drain conductance versus drain-source voltage for different V_{gs} in DMG-GC-DOTTDCD structure.

3. Conclusion

An innovative MOSFET configuration, the dual-material gate graded channel and dual oxide thickness cylindrical gate with two dielectric constant different (DMG-GC-DOTTDCD) MOSFET, has been thoroughly examined by developing analytical models to assess channel electrostatic potential and subthreshold current behaviors. The analytical subthreshold behavior models reveal that the DMG-GC-DOTTDCD MOSFET exhibits superior resilience against short-channel effects (SCEs) in comparison to the DMG-GC-DOT MOSFET.

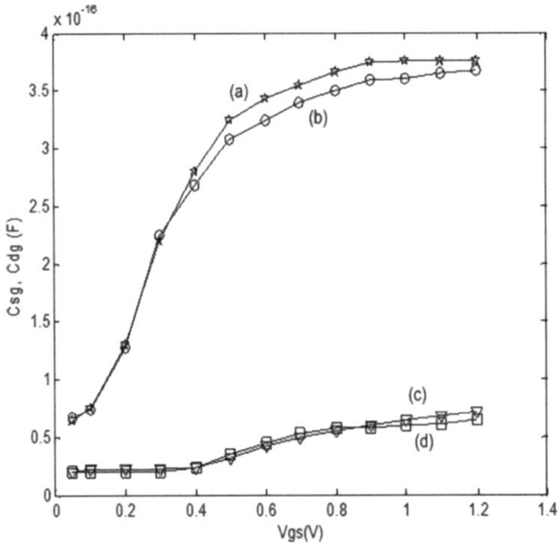

Figure 12.
Capacitance values for drain-to-gate (c, d) and source-to-gate (a, b) are plotted against the gate voltage separately for V_{ds} = 0.1 V (a and c) and V_{ds} = 1 V (b and d).

The incorporation of two oxide thicknesses with varying permittivity has proven effective in enhancing the subthreshold current performance of the DMG-GC-DOTTDCD MOSFET. The analytical models' results align well with the outcomes obtained from numerical simulations conducted using the TCAD simulator.

Author details

Hind Jaafar[1]*, Abdellah Aouaj[1], Ahmed Bouziane[1] and Benjamin Iñiguez[2]

1 Faculty of Sciences and Technology, Laboratory of Automatic, Energy Conversion and Microelectronics (LACEM), Microelectronics and Nanotechnology, Sultan Moulay Slimane University, Béni Mellal, Morocco

2 Department of Electronic, Electrical and Automatic Control Engineering, Universitat Rovira i Virgili, Spain

*Address all correspondence to: jaafarhind@gmail.com

References

[1] Pratap V, Ghosh P, Haldar S, Gupta RS, Gupta M. An analytical subthreshold current modeling of cylindrical gate all around (CGAA) MOSFET incorporating the influence of device design engineering. MJ. 2014;**45**: 408-415

[2] Li C, Zhuang Y, Han R. Cylindrical surrounding-gate MOSFETs with electrically induced source/drain extension. MJ. 2011;**42**:341-346

[3] Cong L, Yi-Qi Z, Li Z, Gang J. Quasi-two-dimensional threshold voltage model for junctionless cylindrical surrounding gate metal-oxide-semiconductor field-effect transistor with dual-material gate. CPB. 2014;**23**:018501

[4] Gautam R, Saxena M, Gupta RS, Gupta M. Two dimensional analytical subthreshold model of nanoscale cylindrical surrounding gate MOSFET including impact of localised charges. JCTN. 2012;**9**:602-610

[5] Dey A, Chakravorty A, Das Gupta N, Das GA. Analytical model of subthreshold current and slope for asymmetric 4-T and 3-T double-gate MOSFETs. IEEE Transactions on Electron Device. 2008;**55**:3442-3449

[6] Zun-Chao L. Dual-material surrounding-gate metal–oxide–semiconductor field effect transistors with asymmetric halo. Chinese Physics Letters. 2009;**26**:018502

[7] Kaur H, Kabra S, Bindra S, Haldar S, Gupta RS. Impact of graded channel (GC) design in fully depleted cylindrical/surrounding gate MOSFET (FD CGT/SGT) for improved short channel immunity and hot carrier reliability. Solid State Electronics. 2007; **51**:398-404

[8] Wu M, Jin X, Kwon HI, Chuai R, Liu X, Lee JH. The optimal design of junctionless transistors with double-gate structure for reducing the effect of band-to-band tunneling. JSTS: Journal of Semiconductor Technology and Science. 2013;**13**:245-251

[9] Aouaj A, Bouziane A, Nouaçry A. Analytical VTH and S models for (DMG–GC–stack) surrounding-gate MOSFET. International Journal of Electronics. 2012;**99**:141-148

[10] Jaafar H, Aouaj A, Bouziane A, Iñiguez B. An analytical drain current model for cylindrical gate DMG-GC-DOT MOSFET. International Journal of Electronics Letters. 2018;**7**:458-472

[11] Dhanaselvam PS, Balamurugan NB. Analytical approach of a nanoscale triple-material surrounding gate (TMSG) MOSFETs for reduced short-channel effects. Microelectronics Journal. 2013;**44**:400-404

[12] Jaafar H, Aouaj A, Bouziane A. Analytical and numerical modeling of Vth and S for new CG MOSFET structure. International Journal of Information Science and Techniques. 2016;**6**:1-11

[13] Taur Y. An analytical solution to a double-gate MOSFET with undoped body. IEEE Electron Device Letters. 2000;**21**:245-247

[14] Chiang TK. A new compact subthreshold behavior model for dual-material surrounding gate (DMSG) MOSFETs. Solid State Electronics. 2009; **53**:490-496

[15] Kaur H, Kabra S, Haldar S, Gupta RS. An analytical drain current model for

graded channel cylindrical/surrounding gate MOSFET. Microelectronics Journal. 2007;**38**:352-359

[16] Jaafar H, Aouaj A, Bouziane A, Iñiguez B. Analytical study of drain current and transconductance for a new cylindrical gate MOSFET structure. In: Proceeding the fourth International Conference on Optimization and Applications (IEEE-ICOA'2018). Mohammedia, Morocco: IEEE-EXPLORE; 2018

[17] Ghosh P, Haldar S, Gupta RS, Gupta M. An analytical drain current model for dual material engineered cylindrical/surrounded gate MOSFET. Microelectronics Journal. 2012;**43**:17-24

[18] Kim CH, Alejandra C-C, Estrada M, Antonio C, Bonnassieux Y, Horowitz G, et al. A compact model for organic field-effect transistors with improved output asymptotic behaviors. IEEE Transactions on Electron Devices. 2012;**60**;1136-1141

[19] Jaafar H, Aouaj A, Bouziane A, Iñiguez B. A compact model of transconductance and drain conductance for DMG-GC-DOT cylindrical gate MOSFET. International Journal of Reconfigurable and Embedded Systems (IJRES). 2020;**9**:34-41

Chapter 3

Electrophysical Properties of Nanofilms Produced by Magnetron Sputtering Method

Dovranov Kuvondik Turakulovich

Abstract

In this chapter, we have measured the formation, surface morphology, elemental analysis, and electrophysical parameters of chromium disilicide thin films and silicon oxide films grown in a magnetron sputtering device by solid-phase ion plasma method. $CrSi_2$ thin films grown by the solid-phase ion-plasma method were initially formed in an amorphous state on silicon and silicon oxide substrates. After thermal annealing at 750 K for 1 h, polycrystalline films were formed as seen from the SEM and RHEED patterns. Electrophysical parameters of silicon and oxide silicon films were measured based on galvanomagnetic phenomena at low temperatures. The optical properties of $CrSi_2$ and SiO_2 thin films were analyzed. The analysis of the wavelength dependence of the light absorption and reflection coefficients of the polycrystalline $CrSi_2$ film showed that it has high sensitivity in the visible and IR regions of light and can be used in nanoelectronic devices operating in these ranges. According to the measurement results, the electrophysical and optical properties of the thin films produced by the solid-phase ion-plasma method in the magnetron sputtering device correspond to the values in the available literature. The solid-phase ion-plasma method in a magnetron sputtering device was found to be highly effective in forming nanofilms.

Keywords: magnetron sputtering, chromium disilicide, solid-phase ion-plasma, electrophysical parameters, reflection coefficient, silicon oxide

1. Introduction

At present, there is growing interest in obtaining nanoscale films and multilayer structures based on dielectric and silicide materials and studying their physicochemical properties. In recent years, scientific research institutes and university scientific laboratories have been focusing on improving environmentally friendly energy sources and finding new energy sources by creating materials with high thermoelectric properties. $CrSi_2$ is a possible convenient material for thermoelectric power generation due to its high efficiency at high temperature. Therefore, the development of technology for synthesis of chromium disilicide and silicon oxide nanoscale films, determination of their structure and properties, and study of mechanisms of formation of new compounds is one of the urgent problems of physical electronics [1–3].

In Central Asia, the leading scientists in creating nanofilms used various methods are M.T. Normuradov, B.E. Umirzakov, and others who conducted scientific research on forming Mn_4Si_7, $BaSi_2$, $CoSi_2$, and CoSi films of various thicknesses and improving their electrical and electronic properties by ion implantation and laser sputtering methods. As a result of these studies, it was found that the increase and decrease of photoluminescence depends on the type of alloying compound, as well as on the postimplantation thermal treatment conditions. However, the optical, electrophysical, electronic, and thermoelectric properties of chromium disilicide and silicon oxide films by the solid-phase ion-plasma method have hardly been studied.

The study of physical processes occurring during the interaction of ions and electrons with various multicomponent materials is of great scientific and practical interest for solving a number of the most important problems of modern electronics. Exposure to ion rays leads to destruction of the structure of samples, the appearance of radiation defects, and the introduction of foreign impurities, which leads to changes in the physicochemical, mechanical, optical, and emission properties of the material irradiated with ions. The creation of new materials with improved physical properties of surface layers, in particular, efficient emitters of secondary electrons based on dielectrics, is one of the urgent problems of microelectronics, since they are widely used in electron beam storage devices and optical devices, magnetrons, etc. One of the methods of directional change in the electrical and other properties of solids is the ion-plasma method, which has a number of advantages over other doping methods. Using low-energy ions, it is possible to control such parameters as the atomic and molecular transfer of oxygen to the silicon surface, the thickness of the doped layer, and the concentration of the introduced mixture, as a result of which the method has become widespread, in semiconductor technology and microelectronics [4].

2. The formation and morphology of nanofilms

Currently, the main idea in the development of ion-plasma methods for deposition of thin films is to lower the operating pressure and voltage across the discharge gap, reducing the energy intensity of the heating processes of the deposition objects. The production processes of microelectronics and nanoelectronics products require the application of multilayer coatings such as semiconductor-dielectric, metal-dielectric-metal, and metal-semiconductor-dielectric-metal, which requires the use of multi-profile vacuum-plasma equipment.

In our studies, to obtain thin films on the surface of silicon single crystals, we used a standard magnetron sputtering installation EPOS-PVD-DESK-PRO, designed for applying one-sided thin film coatings (metals (Ti, Zr, Al, Cu, Au, Ag, Cr, etc.) and resistive layers) by magnetron heating of substrates [5–7]. Using magnetron sputtering systems, tape conductors of transition metals and alloys are metalized, films of refractory materials (molybdenum, titanium, tungsten, etc.), films of nitrides, oxides are obtained, coatings are applied to dielectrics, metals, semiconductors, and coatings are obtained over large areas.

The principle of operation of magnetron-type sputtering devices is based on the use of crossed electric and magnetic fields to form the trajectory of electrons in the plasma in the form of a closed spiral curve.

The magnetic field, the lines of force of which are parallel to the surface of the magnetron sputtering target, holds electrons in close proximity to the target, in the so-called electron "trap" created by crossed electric and magnetic fields. Oscillations of electrons and their movement along spiral trajectories increases the number of ionization collisions, resulting in an increase in plasma density, the effect of which on the target ensures a high sputtering rate of the target material.

The magnetron sputtering system designed for the formation of nanofilms has the following characteristics: (1) vacuum level 10^{-6} mmHg; (2) target diameter 7.6 cm; (3) sputtering speed (depending on the type of sprayed materials) up to 7.8 Å/sec; (4) film thickness measurement accuracy 0.15 Å; (5) adjustable output power from 0.15 to 3.5 kW; (6) substrate position accuracy up to 0.5 mm; (7) coating thickness from tens of nanometers to several microns; and (8) adjustable distance of the target relative to the magnetron up to 160 mm.

The synthesis of thin films formed in a magnetron sputtering installation in the research laboratory "Micro and Nanoelectronics" of Karshi State University is carried out using SiO_2/Si and Si (111) substrates. Before forming the $CrSi_2$ film, the SiO_2/Si and Si (111) substrates were cleaned in two stages:

- Cleaning the surface of Si (111) and SiO_2/Si (d = 60 mm) substrates was carried out by washing with deionized water at a temperature of 65–75°C using an ammonia-peroxide mixture and drying in a centrifuge;

- Vacuum treatment (cleaning) of the surface of the silicon wafer was carried out using a flow of argon plasma produced on an improved EPOS-PVD-DESK-PRO magnetron sputtering. Plasma current is created by a cold cathode ion source with a voltage of 2.5–3.5 kV and a current of up to 125 mA for 5–8 minutes. During the spraying process, a group of pallets (1÷3 pieces) is placed in a rotating device. $CrSi_2$ films were formed on an EPOS-PVD-DESK-PRO magnetron sputtering installation at a pressure of 10^{-6} Pa and at room temperature. The target purity of $CrSi_2$ was 99.6%. The diameter and thickness of the target were 76 and 6 mm, respectively.

Before placing the target in a magnetron machine, its composition and structure were studied using a 3D scanning electron microscope (SEM) Quanta 200 from the Dutch company FEI (**Figure 1**).

Elemental analysis of thin $CrSi_2$ films formed in an automated and modernized magnetron sputtering installation was carried out by energy-dispersive X-ray spectroscopy (EDX), the morphology and surface structure of the films was carried out by high-energy electron diffraction (HEED), the thickness was determined using a scanning electron microscope (standard FEI Quanta installation 200 F).

All $CrSi_2$ films were obtained at room temperature and were amorphous. The films were annealed at 750 K for 2 hours to obtain a polycrystalline structure.

Figure 2 shows an SEM image of an unheated $CrSi_2$ film. The experiment was carried out on a Quanta 200 3D microscope. **Table 1** shows the percentage of elements in $CrSi_2$. As can be seen from the data presented, the sample contains the most chromium and silicon by mass and atomic percentage and very little oxygen.

The table shows that oxygen is present in the smallest quantity. However, a film produced by magnetron sputtering cannot contain C and O atoms. So, these may be environmental atoms located on the surface of the film. A fragment of the surface was taken to study the distribution of elements in depth. The selected fragment represents the surface of the sample at an angle of 48 degrees. Thanks to this, a picture of the distribution of elements along the depth of the sample was obtained (**Figure 3**).

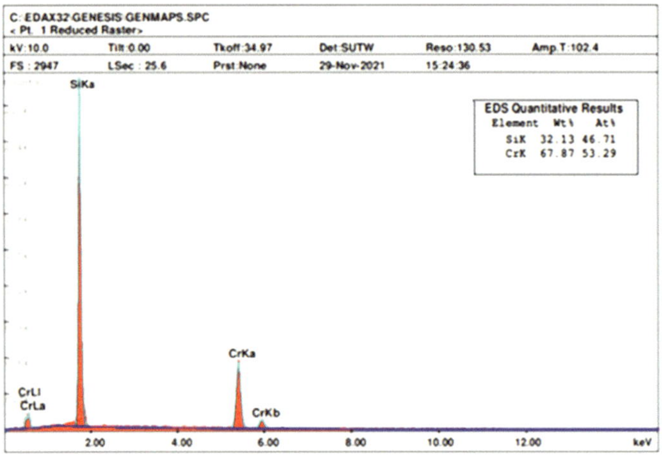

Figure 1.
Results of studying the $CrSi_2$ target using a scanning electron microscope (SEM): Surface image and energy-dispersive X-ray spectrum.

Figure 2.
SEM image of an arbitrary sample surface for elemental analysis.

Energy dispersive spectrum quantitative results		
Element	**Mass**	**at. %**
CrK	36,35	33,6
SiK	60,4	62,32
OK	3,25	4,08

Table 1.
The elemental composition of the sample.

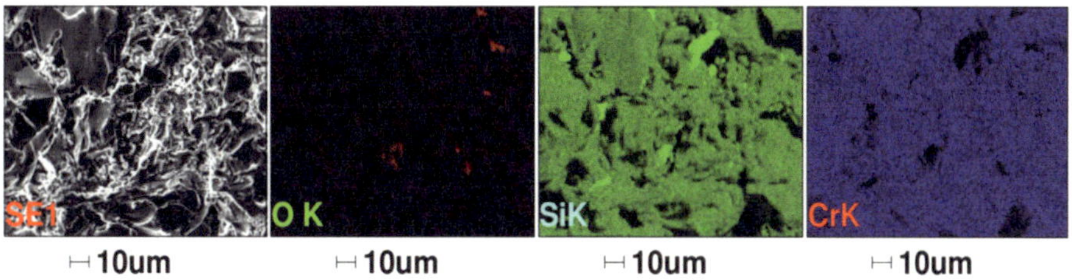

Figure 3.
Map of the elemental composition of samples.

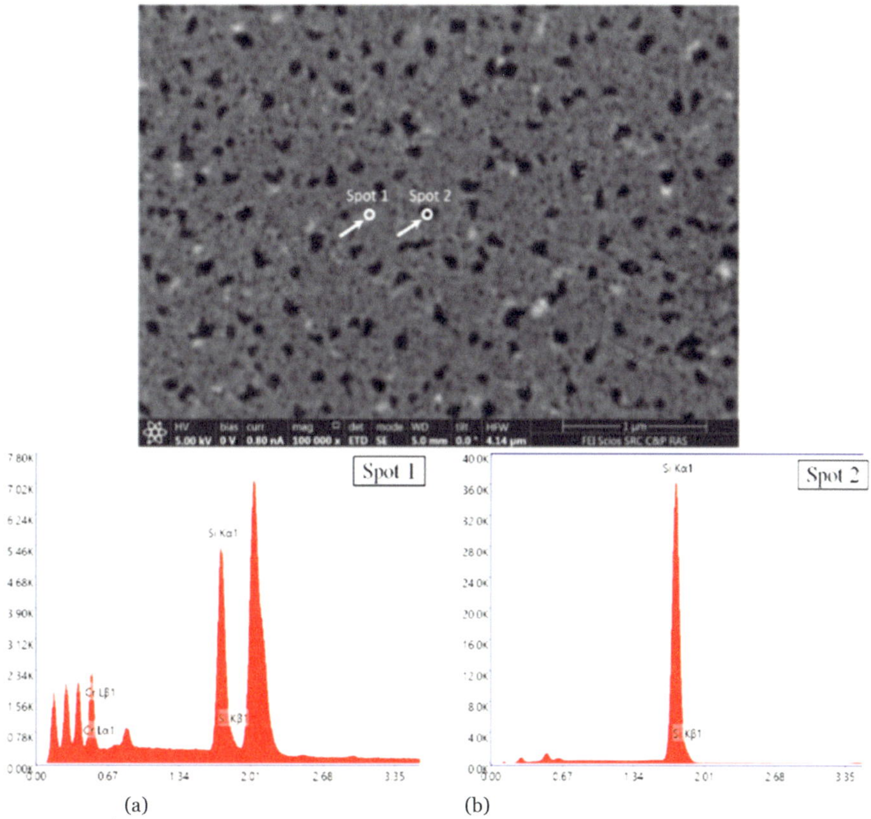

Figure 4.
SEM image (a) and energy-dispersive spectra (b) of the surface of the $CrSi_2$/Si (111) film, deposition time of 30 seconds.

The SEM image (a) and energy-dispersive spectra (b) of the surface of $CrSi_2$ films obtained by the solid-phase ion-plasma method are presented in **Figure 4**. The films were obtained by sputtering onto the surface of a silicon substrate for 30 seconds. As can be seen from **Figure 4a**, individual sections of the film measuring 90–150 nm (black phases) are not coated with silicone. Analysis of the energy-dispersive spectra of the $CrSi_2$ surface (**Figure 4b**) shows that the concentration of Cr and Si at point 1 is about 33.6 and 62.32 at. %, which corresponds to the structure of $CrSi_2$. Point 2 has a very low concentration of Cr atoms. It can be seen that a certain number of $CrSi_2$ molecules are also present at uncoated Si sites. The Si (111) surface is completely covered with an amorphous $CrSi_2$ film during sputtering within 60 s (~400 Å). The SEM image and DBED pattern (inset) of a thin $CrSi_2$ film formed on a silicon surface by magnetron sputtering for 120 s and then annealed at T = 750 K for 1 hour are shown in **Figure 5**. As can be seen from the figure, in this case, a smooth and uniform polycrystalline $CrSi_2$ film is formed. **Figure 6** shows an SEM image of a cross-section of a thin film of the $CrSi_2$/Si (111) system. It can be seen that the film thickness is about 80 nm.

In this section, the surface morphology and composition of $CrSi_2$ films formed on the silicon surface were studied using a Quanta 200 3D scanning electron microscope (SEM). $CrSi_2$/Si (111) films were formed from a $CrSi_2$ target using a modern EPOS-PVD-DESK-PRO sputtering magnetron. The composition of the resulting thin film of silicide $CrSi_2$ and the chemical bonds formed in it have been studied.

Analysis of the research results shows that the $CrSi_2$ thin film can be widely used as a semiconductor element to replace materials with n-type physical properties, as well as active parts of modern micro- and nanoelectronic devices in the future.

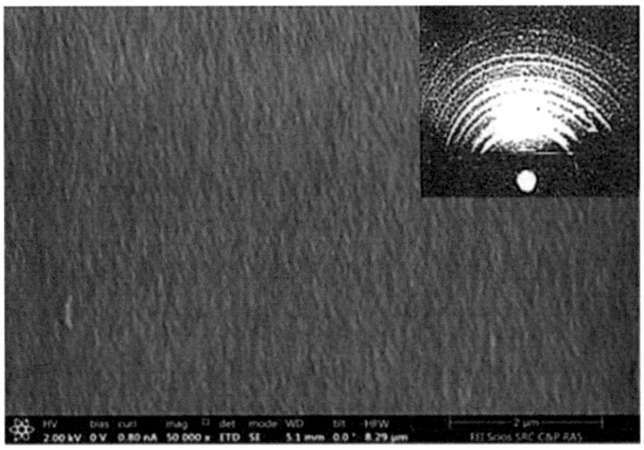

Figure 5.
SEM image and DBED pattern (inset) of a thin $CrSi_2$ film formed by magnetron sputtering for 120 s onto a silicon surface annealed at T = 750 K for 1 hour.

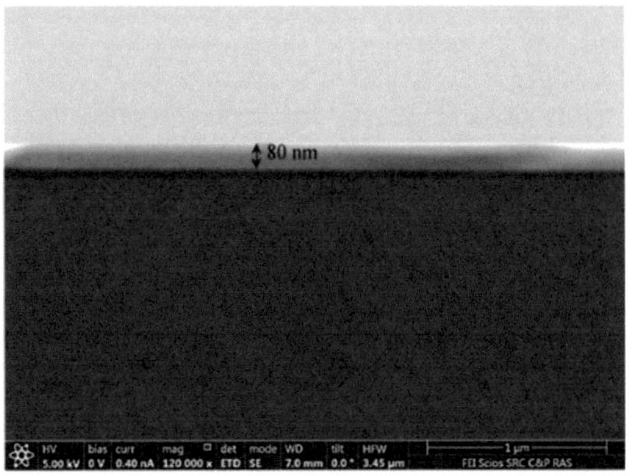

Figure 6.
SEM image of the cross-section of a thin $CrSi_2$ film formed on a silicon surface by magnetron sputtering for 120 s and then annealed at T = 750 K for 1 hour.

3. Band gap energy of thin films using IR and UV spectroscopic methods

The IR spectroscopy method makes it possible to carry out measurements to determine the structural and optical properties of nanomaterials in the solid state. Infrared Fourier spectrometers have significant advantages in spectral resolution, speed, brightness, and therefore, information content compared to spectral instruments that implement traditional schemes for recording optical spectra using dispersive elements. FT-IR spectrometers are used to solve various problems of modern analysis. Designed for qualitative analysis of thin films and thickness of thin films. It is also possible to create near-infrared photodetectors based on chromium disilicide films. However, growing silicon on $CrSi_2$ films and nanocrystals (NC) to create such devices integrated on a silicon substrate based on silicon planar technology remains unexplored. In this work, we investigated the effect of a preformed copper monolayer phase on a monolayer of films in a chromium disilicide system, as well as the effect of chromium disilicide layer thickness on the properties of two different Si film heterostructures by ion-plasma sputtering.

The Happ-Ganzel method was used in the IRTracer-100 Fourier transform spectrophotometer. A graph of the dependence of radiation intensity on wave number was obtained.

The FT-IR spectrophotometer is a reliable, compact, and easy-to-use instrument that performs qualitative and quantitative analysis of a wide range of gases, solid samples, and liquids. The Michelson interferometer has a dynamic digital element alignment system with a signal-to-noise ratio of 30,000:1 (for KRS-5, 4 cm^{-1}, 1 min, 2100 cm^{-1}). Rapid LabSolutions IR software, including specialized programs "Analysis of mixtures," "Identification of substances," and enriched with data analysis and processing systems.

In most atoms and molecules, photons of visible light and ultraviolet light have enough energy to cause transitions between different electronic energy levels. From the wavelength of the absorbed light, the energy required to move an electron from a lower energy level to a higher energy level can be determined. **Figure 7** shows a diagram of electronic transitions in formaldehyde and the wavelengths of light that cause them.

As a result of these transitions, very narrow absorption bands appear at wavelengths that are very characteristic of differences in the energy levels of the absorbing particles. This is also true for atoms, as shown in **Figure 7**. Incident light of a certain wavelength causes the electrons in the atom to be excited.

The type of atom or ion (that is, what element it is) and the energy levels moving between electrons determine the wavelength of light that is absorbed. Transitions can occur between multiple energy levels, with more energy, that is, lower wavelengths of light, required to move an electron further from the nucleus. However, for molecules, vibrational and rotational energy levels are added to the electronic energy levels. Since many transitions with different energies can occur, the bands broaden (**Figure 8**). In solutions, the expansion is even greater due to solvent-solvent interaction. Most nanofilms studied with UV-Vis spectrophotometers have natural broad peaks of 20 nm or more on the side. A UV-Vis device is defined as the width of the UV line at half maximum and is commonly referred to as full width at half maximum (**Figure 9**).

The optical bandgap is an important factor for semiconductors. The term band gap refers to the energy difference between the top of the valence band and the bottom of

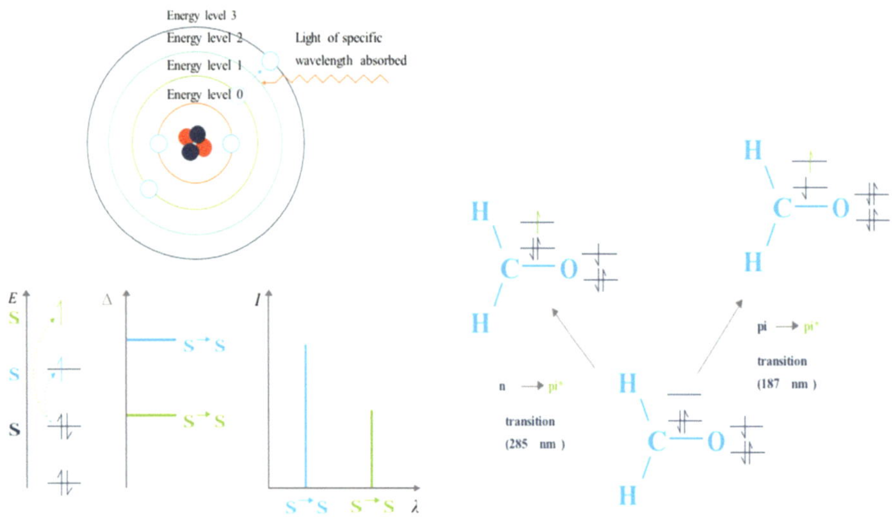

Figure 7.
Electronic transitions in formaldehyde.

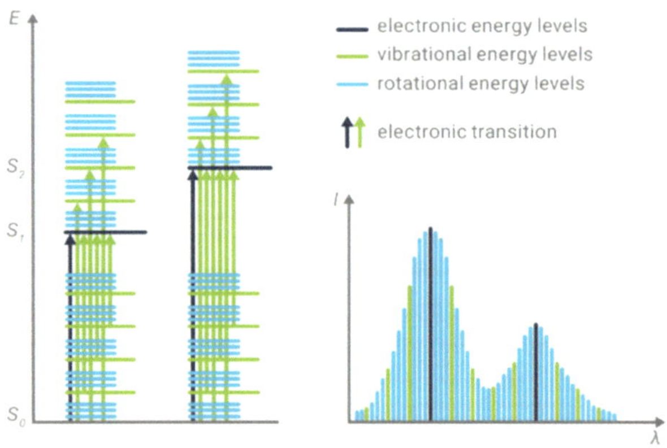

Figure 8.
Electronic transitions and ultraviolet–visible spectra in molecules (I – Intensity and λ – Wavelength).

the conduction band, where electrons can jump from the valence band to the conduction band. In order for an electron to move from the valence band to the conduction band, it needs a certain minimum energy [8]. A bandgap can be classified directly or indirectly if the energy minimum (lower part) of the conduction band is directly above the energy maximum (upper part) of the valence band. If the crystal pulse (k-vector) has the same bandwidth is called "direct." Otherwise, it is called "indirect" bandwidth [9].

A UV spectrometer is used to measure the optical properties of a sample. It consists of a spectral range light source and a fixture for measuring KBr visible and IR transmittance, with a monochromator to select one frequency (wavelength) from all frequencies provided by the radiation source to scan at the desired frequency. For single-crystalline materials, diffuse reflection is an excellent method for analyzing the optical properties of a sample [10].

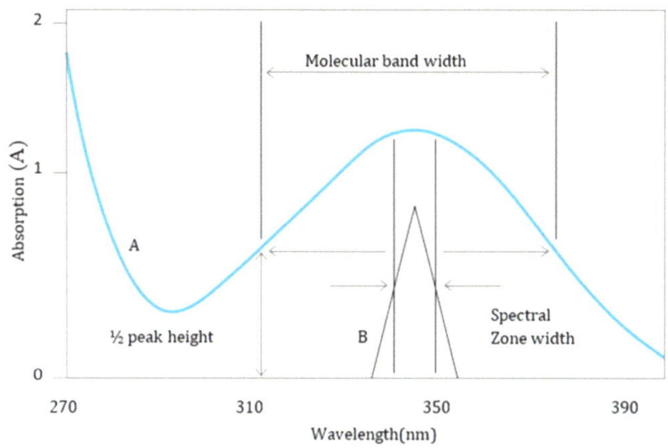

Figure 9.
Spectrum a corresponds to the maximum peak at a wavelength of 345 nm. The wavelength dependence of the absorption (spectral bandwidth) is shown.

When electromagnetic radiation is directed onto the surface of a thin film, two types of reflection occur: specular reflection and diffuse reflection (DR). Specular reflection is direct radiation from the film surface, corresponding to Snell's law of reflection; the angle of incidence is equal to the angle of reflection. Diffuse reflection (DR) is the wavelength-dependent absorption of radiation incident and scattered on a film and within a nanomaterial. Some of this radiation eventually leaves the bulk sample in all directions. The DR accessory is not a mirror component but is designed to collect diffuse reflection and direct it to a photodetector. Measuring diffuse reflectance (DR) using a UV-Vis spectrophotometer is a standard method for determining the optical properties of thin films [11]. In the case of $CrSi_2$ semiconductor nanomaterials, these properties are potentially dependent on the absorption coefficient (K) and band gap energy (E_g). In particular, E_g is an important property of semiconductor nanofilms that determines their applicability in optoelectronic applications. In DR spectroscopy, there are many methods for producing thin films [12].

Using a molecular turbopump (Pfeiffer vacuum), a high vacuum was created in the chamber of the EPOS-PVD-DESK-PRO magnetron device at a pressure of 10^{-6} Torr. First, the surface of the clear glass and silicon wafers was mechanically cleaned with 30% dilute hydrofluoric acid, then the silicon surface was re-cleaned by ionizing argon gases inside the chamber using an installed "ion gun" treatment device in the magnetron chamber. For the first time, we have obtained thin films of $CrSi_2$/Si (111) on heated substrates in a vacuum of $1.2 \cdot 10^{-4}$ mbar using the solid-phase "ion-plasma" method [13].

Figure 10 shows a 3D image of a chromium disilicide nanofilm grown by magnetron sputtering, obtained using an atomic force microscope. The smallest thickness of the $CrSi_2$ film is 80 nm, and the surface is completely covered by the peak. **Figure 11** shows a 2D image of a $CrSi_2$ nanofilm. Moreover, if we take images in only two dimensions, we see that we cannot easily discern differences in surfaces. The height of the islands is more visible in 3D. In both cases, evenly distributed waves are generated under the surface over large areas. This means that the thin surface layer is chromium disilicide. AFM analysis shows that evidence of formation is a nano-sized film obtained by the ion-plasma method.

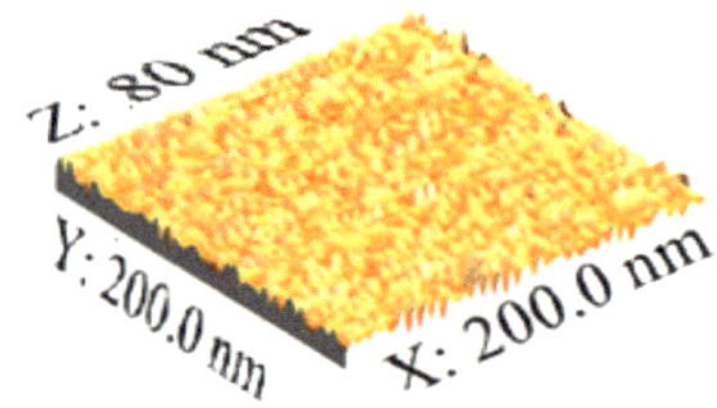

Figure 10.
Three-dimensional images of the surface of thin $CrSi_2$ films obtained by atomic force microscopy.

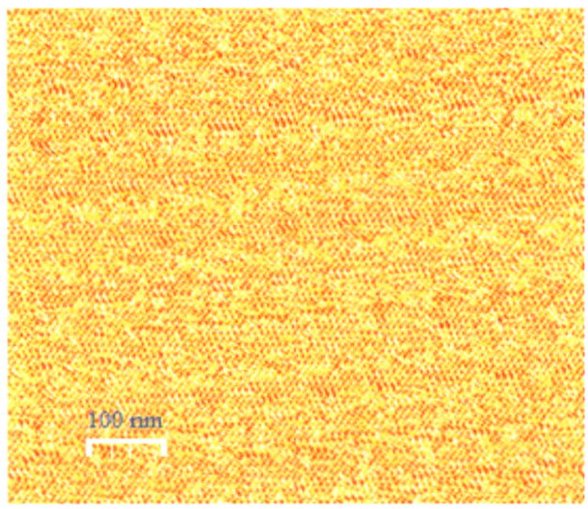

Figure 11.
Two-dimensional images of the surface of thin $CrSi_2$ films obtained by atomic force microscopy.

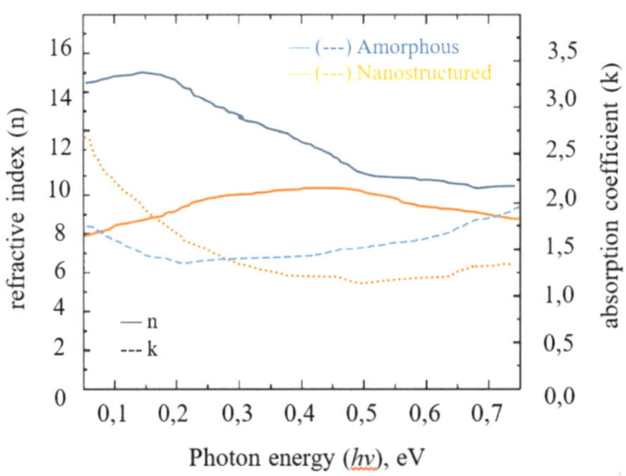

Figure 12.
Refractive index and absorption coefficient of films depending on photon energy: – (---) amorphous (sprayed) and – (---) nanostructured.

In order to measure the forbidden zone energy of chromium disilicide films obtained by the solid-phase ion-plasma method using IR absorption and UV spectroscopy, the dependence of the ultraviolet absorption coefficient and the refractive index of the films on the energy of the incident photons was measured (**Figure 12**).

In this paper, the bandgap width of semiconductor films was calculated using absorption data obtained by spectroscopic methods using the Kumar, Tauts, and

Kubelka-Munk models. The best fit (correlation coefficient was 0.986) was obtained for $CrSi_2$ with values of n = 2 and n = 0.5. **Figure 13** shows the spectral dependencies of absorption in the coordinates of optical transitions based on the Kubelka-Munk and Kumar approach. Based on these correlations, band gap (optical band gap) values were obtained. Approximating the linear range of the spectral dependence with a straight line up to the x-axis intersection gives values of 1.1512 and 0.3543 eV for Si (111) and $CrSi_2$/Si (111) thin films, respectively.

The purpose of this work is to study the morphology, elemental composition, and optical and electronic characteristics of thin $CrSi_2$ films obtained by the solid-phase ion-plasma method. The obtained values of the energy gap and phonon energy for thin films of Si (111) and $CrSi_2$/Si (111) are summarized in **Table 2**.

The results of spectroscopic studies illustrating the absorption spectra for Si (111) and Si (111) with a $CrSi_2$ film 80 nm thick are presented in **Figure 14**. The curves consist of two parts. In the first section, as $h\nu$ increases, the value of I practically does not change, that is, no absorption is observed, most of the photons pass through the sample, and some are reflected. It can be seen that in the case of pure silicon $h\nu \approx 0.22 \div 0.93$ eV, the transmission coefficient T ≈ 0.84, and the reflection coefficient R ~ 0.2. In the case of the $CrSi_2$/Si (111) film in the range $h\nu \approx$ 0.23 eV, T ≈ 0.64 and R ≈ 0.32. In the second part of the $I(h\nu)$ dependence, I drops sharply to zero. From the edge of the absorption spectrum, by extrapolating part of the curve onto the $h\nu$ axis, the approximate value of the band gap is determined. It can be seen that E_g ~ 1.14 eV for Si (111) and E_g ~ 0.34 eV for $CrSi_2$ at room temperature.

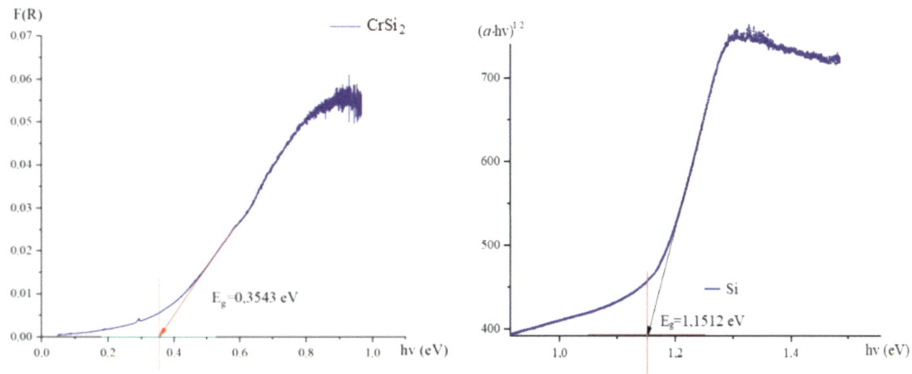

Figure 13.
Tauc plot for $CrSi_2$ and Si thin films for indirect transition according to the Kubelka-Munk model.

Sample Methods	In this work		In literature	
	Si (111) E_g (eV)	$CrSi_2$ E_g (eV)	Si (111) E_g (eV)	$CrSi_2$ E_g (eV)
Kubelka-Munk	11,512	0,3543	1,1	0,34
Kumar	1,132	0,362	1,09	—
Tauca	1,14	0,34	1,13	0,36

Table 2.
The bandgap values of thin films are determined by various methods.

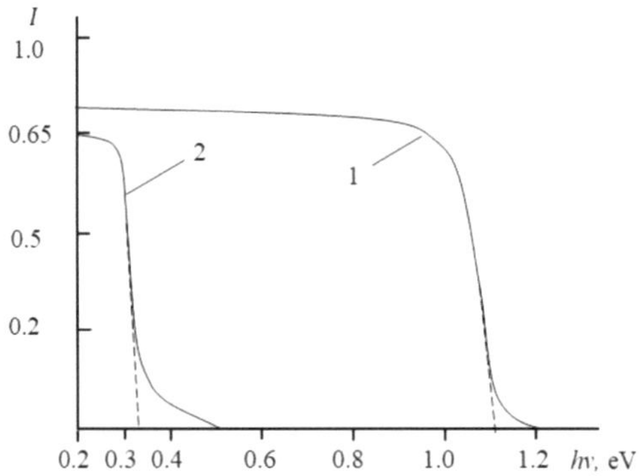

Figure 14.
Light absorption spectra: 1 – Pure Si (111), 2 – $CrSi_2$/Si (111).

4. Formation of silicon oxide films in a magnetron device using low-energy ions

The study of physical processes occurring during the interaction of ions and electrons with various multicomponent materials is of great scientific and practical interest in solving a number of the most important problems of modern electronics. Exposure to ion rays leads to the destruction of the structure of samples, the appearance of radiation defects, and the introduction of foreign impurities, which leads to changes in the physicochemical, mechanical, optical, and emission properties of the material irradiated with ions. The creation of new materials with improved physical properties of surface layers, in particular, efficient emitters of secondary electrons based on dielectrics, is one of the urgent problems of microelectronics since they are widely used in electron beam storage devices and optical devices, magnetrons, etc. One of the methods for directional change in the electrical and other properties of solids is the ion-plasma method, which has a number of advantages over other doping methods. Using low-energy ions, it is possible to control such parameters as the atomic and molecular transfer of oxygen to the silicon surface, the thickness of the doped layer, and the concentration of the introduced mixture, as a result of which the method has become widespread in semiconductor technology and microelectronics.

Amorphous porous dielectric films are promising materials for micro-, nano-, and optoelectronics. These materials are used in LEDs, photodetectors, vacuum microelectronic cathodes, biological implants, gas sensors, and membranes. Based on them, it is possible to manufacture unheated electronic sources and memory elements. One such material is porous silica. It holds great promise for creating humidity, gas, chemical, and biological sensors, as well as other applications. The production and study of amorphous porous films is a priority direction in physical electronics [14].

The study of dielectric films before and after ion doping is of scientific interest since it allows us to understand the processes occurring in dielectrics during ion bombardment, changes in the conditions for the formation and release of secondary electrons, and also explain the nature of the high efficiency of secondary emission of complex dielectrics. In this work, dielectric SiO_2 films obtained by various oxidation methods were studied. There are several ways to produce silicon dioxide: thermal oxidation and plasma-chemical deposition [15], in atmospheric pressure water vapor, ion plating [16], by implanting low-energy $^{+}O_2$ ions in a magnetron device and by the ion implanting [17] method using a magnetron sputtering system. Experiments have shown that SiO_2 films obtained in water vapor are less perfect, while films obtained in an atmosphere of dry oxygen are oxidized evenly. Good adhesion, absence of defects, and smooth surface are characteristic of SiO_2 films obtained by placing $^{+}O_2$ ions on silicon. In addition, it should be noted that the ion plasma method has advantages over other production methods since it allows one to obtain films of a certain thickness, from 8 nm to several microns, with a high level of chemical purity [18].

The use of dielectric materials, increasing demands on the quality of coatings, and increasing substrate sizes have led to changes in magnetron sputtering systems using high-frequency voltages.

In the chamber of the EPOS-PVD-DESK-PRO magnetron device, a high vacuum was created at a pressure of 10^{-6} Torr using a molecular turbopump. By introducing argon into the chamber, a pressure sufficient to create the ion-plasma method was created, and this process was controlled on the CP-307 monitor of the magnetron device. Initially, thin films of silicon oxide of various thicknesses were created using the ion-plasma method at room temperature and pressure of $1{,}3 \times 10^{-5}$ mbar.

SiO_2 films are deposited on polished single-crystalline silicon substrates. During the experiments, the substrates were mounted on a rotating carousel holder at a distance of 150 mm from the surface of the magnetron target.

The formation of a thin SiO_2 film on the silicon surface was carried out in the following stages:

- The surface of the substrate was previously mechanically cleaned with 30% diluted hydrofluoric acid;
- Plasma current in a vacuum is created by an ion source with a cold cathode with a voltage of $1.5 \div 3.5$ V and a current of up to 100 mA for $4 \div 6$ minutes. To improve the covalent bonding of Si-O atoms, the surface of the base silicon was repeatedly cleaned with Ar^{+} gases using an "ion gun";
- $^{+}O_2$ was sprayed onto the silicon surface and cleaned with low-energy ions in various doses through a magnetron sputtering device;
- In the modernized Epos-PVD-Desk-Pro device, to obtain a high-quality silicon oxide film, the base was simultaneously heated to a temperature of 400 K using a special heater.

The experiment involved thin films of SiO_2 formed in an improved magnetron sputtering device. The sample was formed on the surface of Si substrates with p-type conductivity and a resistivity of 1.2 Ω ×cm. First, the silicon surface was cleaned with 30% hydrochloric acid. Thermal heating to a temperature of 450 K in an automated furnace was controlled through a magnetron device monitor. In a vacuum of 10^{-5} mmHg, the surface was cleaned of various impurities using Ar^{+} gases through an ion gun. Thin SiO_2 films formed on the silicon surface by the ion plasma method using low-energy $^{+}O_2$ ions had a thickness of 0.17 μm (**Figure 15**). The rate of oxygen deposition on the silicon substrate was controlled by a quartz sensor.

SEM image of a thin film of silicon oxide (SiO_2) obtained by a low-energy ion plasma method using a magnetron sputtering device, as well as the chemical composition and energy dispersive X-ray diffraction pattern of a SiO_2 sample obtained after growing oxide layers 150–250 nm thick at high temperature as a substrate, prepared using an energy dispersive device. Analyzes were carried out under a Quanta 200-3D

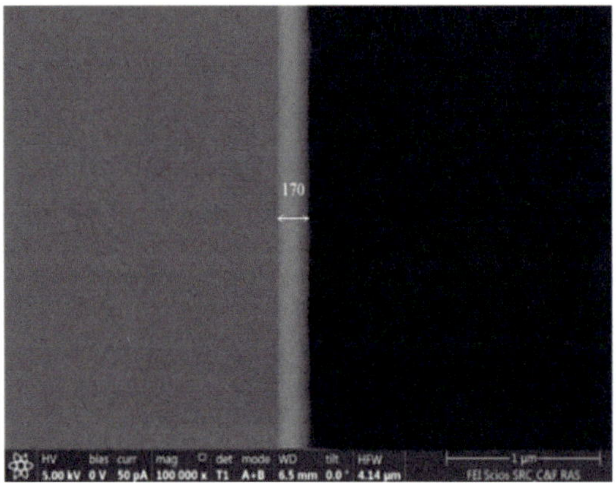

Figure 15.
SEM image of a SiO_2 film formed on a silicon surface.

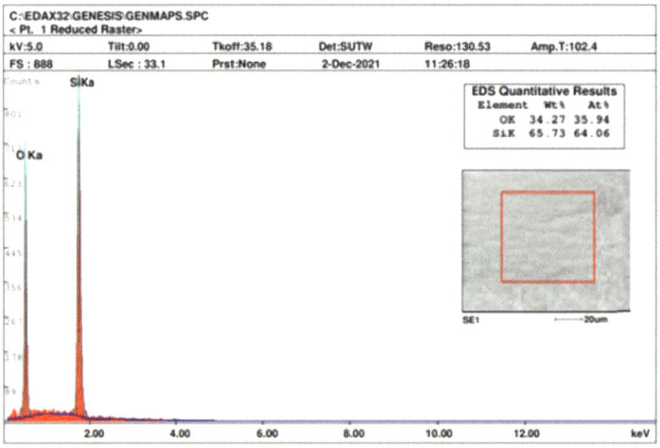

Figure 16.
Chemical composition and energy dispersive X-ray diffraction analysis of SiO_2 film.

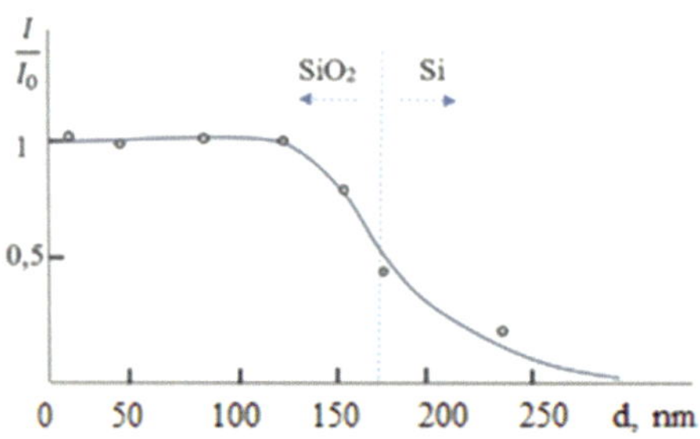

Figure 17.
Diffusion profiles of oxygen atoms along the depth of the SiO_2/Si (111) film.

microscope (**Figure 16**). Results (O-wt.% - 34.27, At% -35.94), (Si-wt.% - 65.73, At% -64.06) elements were determined in fractions.

Figure 17 shows the profile of the distribution of oxygen atoms over the depth of a continuous SiO_2/Si (111) film with a thickness of ≈170 nm, formed by the

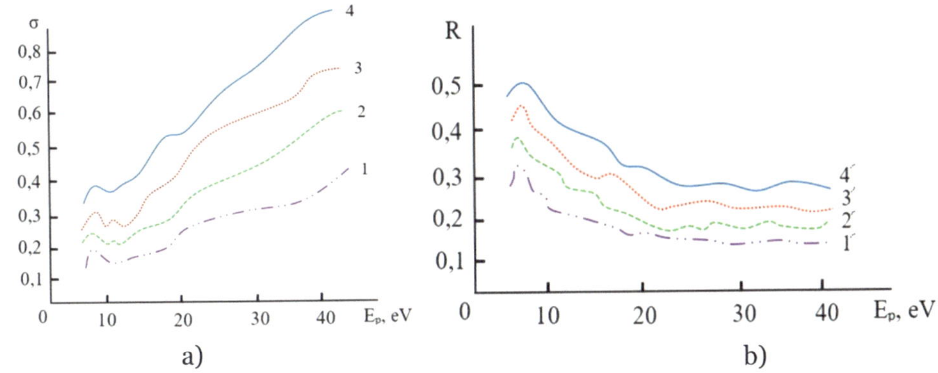

Figure 18.
Dependences of a) – σ (1–4) and b) – R (1′-4′) on ep in the low energy region for the obtained SiO_2 films: –1.1′ in water vapor; in dry oxygen - 2.2′; implantation of $^+O_2$ ions into silicon - 3.3′ and deposition of $^+O_2$ ions onto the silicon surface using the ion-plasma method - 4.4′.

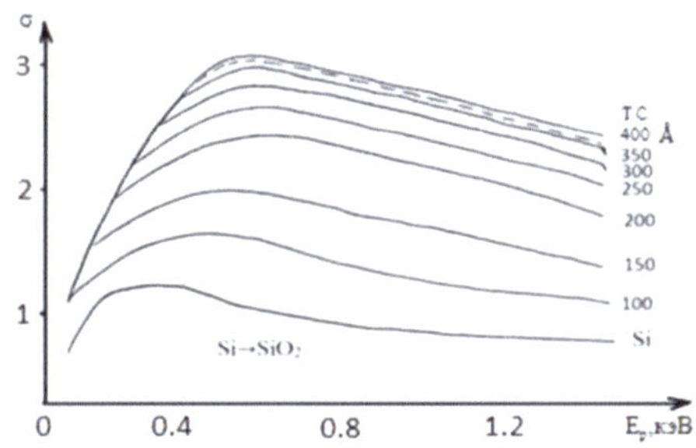

Figure 19.
Dependences of $\sigma(E_p)$ for the Si – SiO_2 system at different film thicknesses.

ion-plasma method with low-energy ions. It can be seen that a transition layer with a thickness of 158 ÷ 182 nm is formed between SiO_2 and Si, in which the oxygen concentration monotonically decreases with depth. The large thickness of the transition layer is apparently associated with a sharp difference in the lattice parameters of Si (c-Si = 5.43 Å) and SiO_2 (α-SiO_2 = 7.13 Å).

It is well known that, compared with metals and semiconductors, the values of the BEE coefficient for dielectrics are mainly due to the large depth of the escape of true secondary electrons (TSE). The latter case is associated with a fairly large band gap (ΔE_g) and a low concentration of free current carriers in dielectrics.

Figure 18 shows the energy dependences of the emission coefficient of secondary electrons σ (E_p) and elastically reflected electrons R(E_p) for the low-energy region of primary electrons E_R obtained from SiO_2 samples obtained by different methods. Although the numerical values for different films are different, the fine structure of the energy dependence curves in a small range of E_p is practically the same.

Therefore, excited electrons with energy less than ΔE_g lose energy only to excite phonons. Therefore, in dielectrics, secondary electrons can travel a much longer distance than in metals and semiconductors until their energy decreases to a value on the order of electron affinity. To determine the maximum depth of HVE release from silicon oxide, the dependences σ(E_p) were studied in a wide range of E_p when SiO_2 films of various thicknesses were deposited on a silicon substrate (**Figure 19**).

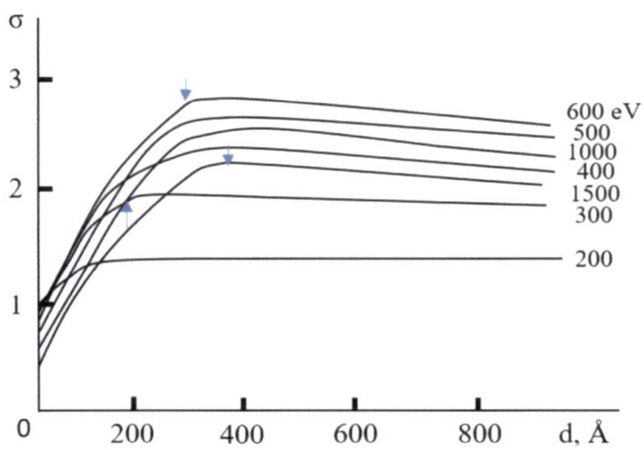

Figure 20.
Dependence of σ on the thickness d of films deposited on the Si surface for different energies of primary electrons.

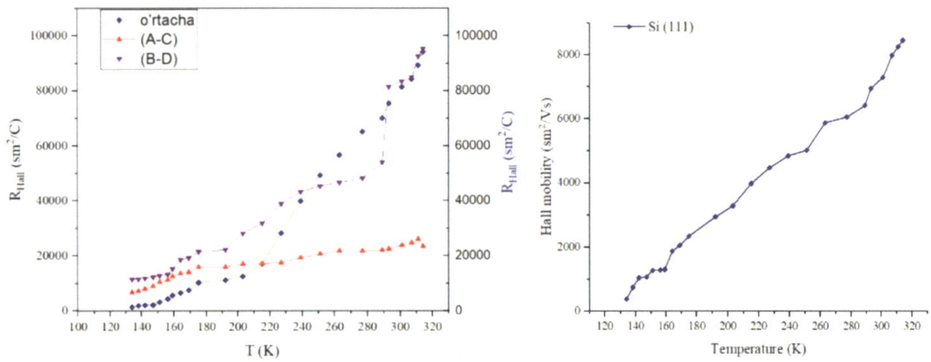

Figure 21.
Temperature dependence of hall mobility and hall constant for silicon.

The film thickness was determined with an accuracy of 10% based on the time the silicone was exposed to an oxygen atmosphere. It can be seen that with increasing film thickness d, the value of the coefficient σ increases and the position of σm shifts toward larger E_p.

This behavior can be explained if we recall that σ reaches its maximum value when the maximum excitation depth of the secondary electrons is equal to their exit zone. At d = 350 Å, based on the practical absence of changes in the position of σm and the shape of the $\sigma(E_p)$ curves, it can be noted that the yield depth for SiO_2 is about 350 Å. These data are presented more clearly in **Figure 20a**, which shows the dependences of σ(d) for various primary electron energies E_p.

It can be seen that at an energy of 600 eV the σ(d) curve reaches a plateau at a film thickness d = 350 Å. For lower energies, the σ(d) curve begins to plateau at small thicknesses, which is explained by the smaller penetration depth of primary electrons at these energies. Thus, from these graphs, it is possible to obtain important information, such as the depth of penetration of primary electrons of different energies into dielectrics, as well as the dependence of the depth of the IHE exit zone on the energy of primary electrons.

This relationship is shown in **Figure 20**. It can be seen that the value of the SPE output depth first increases linearly with increasing E_r, and at 600 eV the dependence $\sigma(E_p)$ reaches a maximum. The value of λ' in the figure corresponds to the maximum value of the depth of the IHE exit zone for pure SiO_2 samples.

5. Electrophysical parameters of thin films

Electrophysical properties of thin-layer silicon oxide films obtained by low-energy (1÷15 eV) ion-plasma method. Low-temperature in-plane electrical properties of SiO_2 samples deposited on a Si substrate in the Van der Pauw configuration using the Hall effect method HMS-5000. Measurement system provided by ECOPIA CORP; measurements were carried out at temperatures from180–27°C. Good electrical connections have been achieved using indium. For electrical measurements at direct current, the width of the square on which the sample with the SiO_2 film was located was 2 × 2 cm, and the size of the sample was 10 × 10 mm. Magnetic field strength up to 0.56 Tesla is created by a superconducting solenoid with an automatic field displacement box. The systems measure the electrical properties of dielectric films (SiO_2) and all semiconductors, including Si, Mn_4Si_7, $CrSi_2$, SiGe, SiC, GaN (can be measured n-type and p-type), metal layers, oxides, and thin films intended for measurement.

Voltages and currents were measured at different points using four probes. AB = 3665.45 mV, BC = 3972.63 mV, CD = 3491.29 mV, DA = 3173.59 mV. For a 190 nm thick silicon oxide film grown on a silicon surface, the Hall coefficient at room temperature is 489.9 cm^3/C. In the following experiments, the low-temperature Hall coefficient of SiO_2 films of various thicknesses was determined. Volume concentration 1.1541×10^{17} cm^{-3}, surface concentration 2.1036×10^{14} cm^{-2}, surface resistance 3146.47 Ω/¤, magnetoresistance 1245.3 Ω, permeability 0.00799 S/cm, and measured vertical/horizontal ratio plane was 0.97469.

It should be noted that the Hall constant for the object under study varies with temperature over a wide range: as the temperature increases from 130 to 300 K, it increases eightfold. As can be seen from the Hall emf measurements (**Figure 21**), the conductivity of silicon used as a substrate is p-type conductivity.

For amorphous SiO_2 films, the volume and surface concentrations at room temperature are equivalent to $N_e = 2.584 \times 10^{15} \pm 0.5 \times 10^{15}$ cm^{-3}, $n_{sh} = 1.44 \times 10^{12} \pm 0.5 \times 10^{12}$ cm^{-2}, mobility $\mu_H = 31.87 \pm 0.04$ $cm^2 \times V^{-1} \times s^{-1}$. As the temperature decreases, the surface concentration decreases. This value is significantly lower than the values reported for amorphous SiO_2 films ($N_e = 4 \div 6 \times 10^{19}$ cm^{-3}) deposited on a silicon wafer. Moreover, it is 2.5 times smaller compared to bulk SiO_2 with $\mu_H = 12$ $cm^2 \times V^{-1} \times s^{-1}$ [19]. Such low mobility can be explained by the structural disorder inherent in amorphous dielectric compounds, which weakly scatter charge carriers. The mobility values vary from $\mu_H = 32.90 \pm 0.08$ $cm^2 \times V^{-1} \times s^{-1}$ to $\mu_H = 36.40 \pm 0.12$ $cm^2 \times V^{-1} \times s^{-1}$ for films with a thickness of 190 nm, respectively. These values remain significantly lower than those measured in the mass analog.

Let us dwell on the results of measuring films of silicon and silicon oxide at low temperatures using the four-probe method on the HMS-5000 device. **Figure 22** shows the temperature dependence of the surface resistance and conductivity of a monocrystalline silicon film intended for use as a substrate. As the temperature decreased, the surface resistance increased and the conductivity decreased, which can be explained by the fact that it is a pure semiconductor film.

Table 3 shows the electrical parameters of thin films of silicon oxide obtained by the ion-plasma method using pure silicon and low-energy ions intended for use as a substrate.

As can be seen from **Figure 23a**, the charge concentration on the surface of thin films of silicon oxide grown on a silicon substrate increases with temperature, and the volume concentration decreases significantly. From **Figure 23b**, it is clear that with increasing temperature, the surface resistance of the film decreases, the concentration

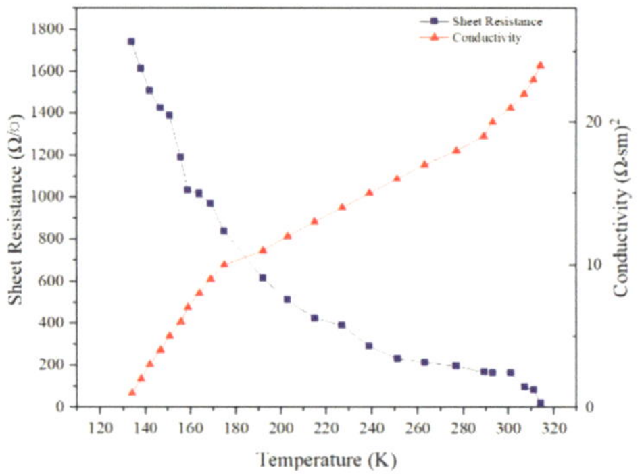

Figure 22.
Temperature dependence of conductivity and surface resistance of pure silicon.

Options	Measurement value	Measurement result for Si substrate	Measurement result for SiO_2 thin film
Bulk concentration (1/cm^3)	$10^7 \div 10^{21}$	$1{,}92{\cdot}10^{14} \div 9024{\cdot}10^{15}$	$1{,}24{\cdot}10^{17} \div 7{,}16{\cdot}10^{15}$
Sheet concentration (1/cm^2)	$10^7 \div 10^{17}$	$11{,}437{\cdot}10^{14} \div 1443{\cdot}10^{12}$	$1{,}33{\cdot}10^{15} \div 1{,}39{\cdot}10^{14}$
Mobility (cm^2/V × sec)	$1 \div 10^7$	984,06 ÷ 2478,3	11,143 ÷ 64,86
Magnetic flux density (T)	0.51Tesla (± 0.03 T)	0,53 Tesla (± 0.03 T)	0,54 Tesla (± 0.02 T)
Temperature (K)	80÷300	130÷300	170÷300

Table 3.
Electrophysical properties of Si and SiO_2 according to the results obtained on the HMS 5000 measuring device.

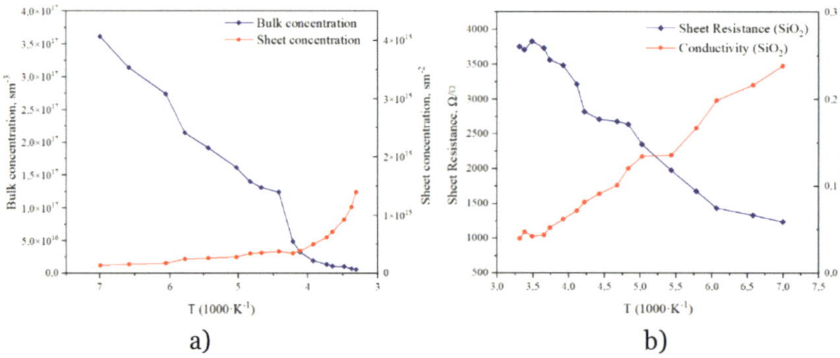

Figure 23.
a) Temperature dependence of the distribution of volume and surface concentration of a thin film of SiO_2, b) temperature gradient dependence of the conductivity and surface resistance of the sample.

of charge carriers on the surface increases, and the conductivity increases to a very small value. The measurement results indicate that SiO_2 films grown by the low-energy ion plasma method have dielectric properties.

The results obtained show that the electrical properties of thin films of silicon dioxide formed by low-energy ions on the silicon surface correspond to literature data. The result of the study shows that a dielectric film is formed.

6. Optical parameters of thin films

The layer thicknesses of thin SiO_2 films grown on the silicon surface were measured by transmission methods and the single reflection ATR method (Ge prism) at incidence angles of 45 and 90°. The thickness of SiO_2 films is 250, 200, and 150 nm. **Figure 24a** shows the ATR measurement results. As the oxide layer becomes thinner, the peaks around 1258 and 1164 cm^{-1} shift toward higher (or lower wave number) wavelengths.

Structural changes in dielectric films were assessed using FT-IR transmission spectroscopy and multiple attenuated total internal reflection (ATR). An IR Fourier spectrometer "IRTracer-100" was used. To study the structure and composition of SiO_2 films by the ATR method, trapezoidal n-germanium samples with a resistivity of 18 $\Omega \times$cm, which are transparent in the wavelength range of interest, were used.

The thicknesses of dielectric films were measured by the ellipsometric method. Scanning electron and atomic force microscopy were used to analyze the effect of high-temperature annealing on the structure of dielectric-semiconductor samples.

Peak height is plotted on the vertical axis and film thickness is plotted on the horizontal axis to determine the relationship between peak intensity and film thickness in the 45° transmittance spectra. **Figure 24b** shows the results for low wavelengths. These results indicate that a good correlation was achieved between peak intensity and film thickness.

They are very similar because they have similar chemical composition and structure. The overall higher intensity in the IR spectra of samples A (green) and B (red) is due to the larger film thickness. The assignment of IR vibration bands to the spectra of films deposited under various plasma conditions is given in **Table 4**.

As can be seen from **Figure 25a**, the IR spectra show characteristic vibrational bands at 1080, 800, and 445 cm^{-1}, corresponding to stretching, bending, and out-of-plane strain of Si—O bonds, respectively. It has a transmittance of 35.21% at wave number 893 cm^{-1} of the main Si-O vibration line and a transmittance of 50% at 758 cm^{-1}. In the absorption spectrum, 45.3% of the light is absorbed at a wavelength of 11197.7 nm. The absorption at 13192.6 nm was 30%, and a well-defined shoulder indicates the stoichiometric structure of silica. At the same time, bands associated with impurities are also observed. The main peaks of light transmission were 893.04 cm^{-1}. Therefore, the existence of stretching vibrations of Si and OH groups at these peaks can be emphasized.

When silicon is oxidized (naturally in the air atmosphere or under the influence of high temperature), hydroxyl groups and SiO_x films are formed on its surface. The

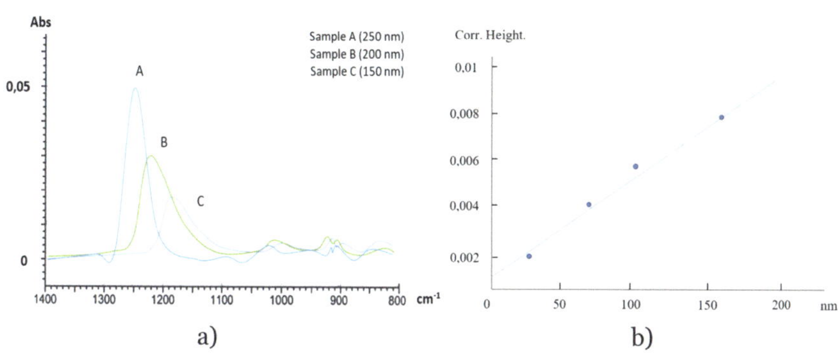

Figure 24.
a) Results of ATR measurements of SiO_2 film, b) photometric graphical results of silicon oxide films of different thicknesses.

Vibrational modes of chemical bonds	Sample A (d = 150 nm) O_2 flow, 3 sccm	Sample B (d = 200 nm) O_2 flow, 10 sccm	Sample C (d = 250 nm) O_2 flow, 14 sccm
Bending vibrations from Si-O plane strain	Large peaks superimposed on the strong band at 445 cm^{-1}	Large peaks superimposed on the strong band at 445 cm^{-1}.	Large peaks superimposed on the strong band at 445 cm^{-1}.
Si-OH stretching Stretching vibrations	760 cm^{-1} weak band	760 cm^{-1} strong band	760 cm^{-1} strong band
Si-O - Stretching vibrations-Si stretching	890 cm^{-1}	895 cm^{-1} with a well-defined step	895 cm^{-1} with a well-defined step
Si-OH deformation; CH bending vibrations	Very small peaks located at 1500–1700 cm^{-1}.	Small peaks located at 1500–1700 cm^{-1}.	Small peaks located at 1500–1700 cm^{-1}.

Table 4.
Results of IR analysis of SiO_2 films of various thicknesses.

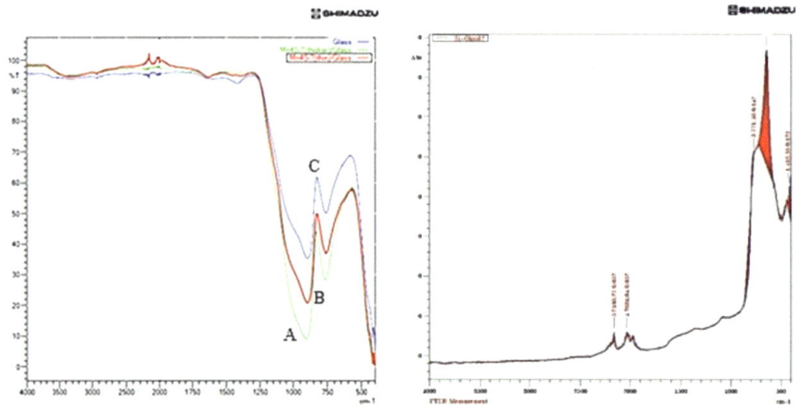

a) b)

Figure 25.
a) Transmission and absorption spectra of thin SiO_2 films of various thicknesses formed by low-energy O_2^+ ions.
b) Smoothing analysis of transmission and absorption spectra of SiO_2/Si.

presence of hydroxyl (Si-OH) groups leads to air humidity, adsorption, and corresponding changes in the spectra of the film on the silicon surface. The absorption and transmission spectra of SiO_2 films obtained by the ion-plasma method were obtained on an IRTracer-100 spectrophotometer using the Happ-Ganzel method. In the range of 400÷4000 cm^{-1}, in order to reduce the influence of water vapor (H_2O) and carbon dioxide (CO_2) molecules, several of the following corrections were made: addition, smoothing, baseline zero correction, normalization, filtering and ATR, and correction.

Figure 25b shows the FTIR spectra of the SiO_2 film formed by low-energy ions on the Si surface. An absorption peak was observed at a wave number of 769.6 cm^{-1}. These peaks correspond to the Si-O-Si group in antisymmetric stretching vibrations, respectively. The wave number of 644.22 cm^{-1} in the absorption spectrum corresponds to the "fingerprint" region of pure Si(111) vibrations.

The modes of vibration of the chemical bond in the silicon oxide film are as follows: -Bending vibrations from Si-O plane deformation have large peaks superimposed on a strong band at a wave number of 445 cm^{-1};

- Si−OH bond stretching vibrations weak band at 760 cm^{-1};
- Si−OH deformation and CH bending vibrations have very small peaks located at 1500–1700 cm^{-1}.

The study of silicon oxides showed that SiO_x (x = 1÷2) is formed during deposition, and with decreasing x, the maximum limit of the lower wave numbers of the n-band during annealing (Si−O−Si) shifts from 915 to 980 cm^{-1} region, on the contrary, increases from 780 to 835 cm^{-1}; the cyclic frequency of oscillation of the pendulum increases with ×.

The penetration depth of infrared light depends on the sample and the refractive index of the crystal. Because the refractive index is wavelength dependent, ATR spectra have slightly different intensity ratios across the spectrum and may require adjustment for comparison with transmission spectra.

We used the following modified Sellmeyer Edwards et al. expression, which is the most appropriate dispersion equation for ambient temperature [20].

$$n^2 = \varepsilon + \frac{A}{\lambda^2} + \frac{B\lambda_1^2}{\left(\lambda^2 - \lambda_1^2\right)} \tag{1}$$

here λ_1 = 11,071 μm, ε = 1.16858:101, A = 939,816·10^{-1} and B = 81,046·10^{-3}. The results obtained gave λ = 199,807 microns for silicon and λ = 10,111,223 microns for silicon oxide.

Using the "film thickness calculation" function in the "LabSolutionsIR" software block of the Fourier-transform infrared spectrophotometer, the following information can be determined for the sample: angle of incidence, refractive index, film thickness, average number of interference fields, and standard deviation. **Table 5** below shows the measured parameters of films of different thicknesses obtained using low-energy ions.

The data in the cited sources is as follows: wavelength 0.5876 μm, refractive index: n = 3.9766, absorption coefficient k = 0.030209, and complex refractive index (n + ik) [21].

Figure 26b shows the spectral dependence of the refractive index of samples A, B, and C, as well as the refractive index of stoichiometric thermally grown SiO_2. The physical composition of the films was analyzed using ellipsometric data. The results showed a slightly lower stoichiometric composition of $SiO_{1.985}$ for sample A and $SiO_{1.938}$ for sample C. However, increasing the oxygen flow rate to 12 cm^3 results in the formation of voids with a volume fraction of ~1.5%. This result explains the observed lower refractive index values with decreasing film thickness.

As can be seen from **Figure 26b**, both the refractive index and extinction coefficient of the SiO_2 film change with wavelength, indicating that the SiO_2 film has a certain dispersion. The refractive index and extinction coefficient of the deposited SiO_2 thin films at a wavelength of 250 nm are 1.494 and 0.00423, respectively.

The obtained results confirm the qualitative and quantitative analysis of the formation of thin layers of silicon oxide grown with the help of low-energy ions by ion-plasma method, SiO_2-Si interface, as well as vibrational methods of chemical bonding, absorption, and transmission spectra of nanofilms. The thickness and refractive index of the thin film were measured by analytical and experimental

Sample	Si (111)	SiO_2/Si (111)	Ref, Si (111)
Range(cm^{-1})	503	465–1397	587,6 [21]
Refractive index	345,323	1494	39,766 [21]
Angle of incidence	45°	90°	
Middle edges of intervention	24	122	
Thickness (μm)	51,34	12,36	
Standard Deviation (μm)	10,06	64,32	
Absorption coefficient	0,03241	0,00423	0,030209 [21]
Minimum Peak (%)	47.6527	64.007	23.47
Maximum Peak (%)	96.5350	92.099	64.35

Table 5.
Comparative comparison of optical parameters of Si and SiO_2 films.

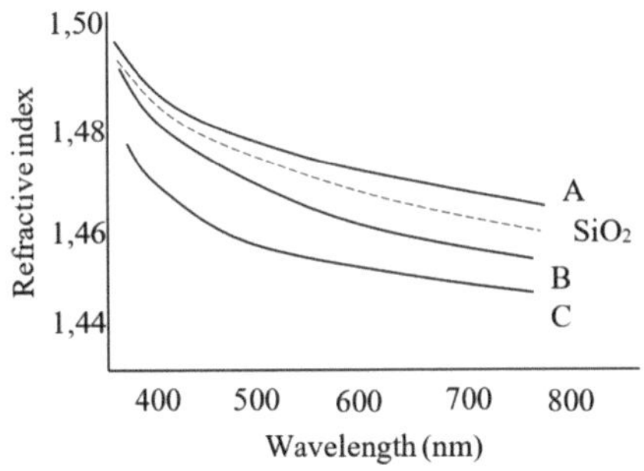

Figure 26.
Refractive index dispersion of silicon oxide films deposited at different thicknesses.

methods. Refractive index for silicon is n = 3.45323. "Mean noise limits" were measured for silicon oxide and pure silicon. The penetration depth of infrared light into the crystal was measured. The results show that the as-deposited SiO_2 thin films can be used in optical films with a low refractive index.

7. Conclusions

The following conclusions were reached on the works produced by the solid-phase ion-plasma method in the magnetron sputtering device:

- For the first time, nanoscale $CrSi_2$ films were formed on the Si (111) surface using a solid-phase ion-plasma method in a magnetron sputtering device. It was found that polycrystalline $CrSi_2$ films were obtained by thermally heating initially amorphous $CrSi_2$ thin films at a temperature of 750 K for 1 hour;

- According to the results of the IR spectrum, it shows characteristic vibrational bands at 1080, 800, and 445 cm^{-1} corresponding to stretching, bending and out-of-plane deformation of Si—O bonds. It has 35.21% transmittance at 893 cm^{-1} wave number of the main Si—O vibrational line and 50% transmittance at

758 cm^{-1}. In the absorption spectrum, 45.3% of light is absorbed at a wavelength of 11197.7 nm. The absorption at 13192.6 nm was 30% and indicated a well-defined stoichiometric structure of silicon. At the same time, the bands associated with the compounds were also observed. The main peaks of transmittance were 893.04 cm^{-1}. Therefore, it can be noted that the stretching vibrations of Si and OH groups are present in these peaks.

- The thickness and refractive index of the thin film were measured by analytical and experimental methods. Refractive index for silicon is n = 3.45323 and for silicon oxide n = 1.494. The penetration depth of infrared light for silicon oxide and pure silicon was found to be 12.34 and 51.36 μm, respectively.

- Mobility values for 170 nm thick SiO_2 films produced by low-energy ions range from $\mu_H = 32.90 \pm 0.08$ cm^2V^{-1} s^{-1} to $\mu_H = 36.40 \pm 0.12$ cm^2V^{-1} s^{-1}, respectively. These values remain significantly lower than those measured in the mass analogue. According to the measurement results, it can be seen that the conductivity and magnetoresistance of the film decreased with decreasing temperature, and it has dielectric properties at room temperature. The mobility of charge carriers, Hall constant, changes in surface and volume concentrations at low temperatures were analyzed. SiO_2 thin films formed by the ion-plasma method can be used as dielectric layers of transistors, modern silicon integrated circuits and solar cells based on their dielectric properties.

Acknowledgements

I am very grateful to Elena Nikolaevna Vlasova, Head of Laboratory No. 21 of the Institute of Macromolecular Compounds of the Russian Academy of Sciences (IMC RAS) and Muradulla Normuradov, Professor of the Department of Theoretical and Experimental Physics of Karshi State University (Uzbekistan), for their advice on this work.

Conflict of interest

"The authors declare no conflict of interest."

Author details

Dovranov Kuvondik Turakulovich
Karshi State University, Karshi City, Uzbekistan

*Address all correspondence to: quvondiqdavronm@gmail.com

References

[1] Timm MM. Ion Implantation and Ion Irradiation Effects on Nuclear and Thermoelectric Materials. Porto Alegre: Universidade federal do rio grande do sul instituto de física programa de pós-graduação em física; 2019. pp. 11-25, 106

[2] Makram AQ, Venkat R, Kumar R, Hartmann T, Ginobbi P, Newman N, et al. Structural, electrical, and thermoelectric properties of $CrSi_2$ thin films. Thin Solid Films. 2013;**545**:100-105. DOI: 10.1149/MA2010-02/19/1374

[3] Galkin NG, Goroshko DL, Galkin KN, et al. Influence of Cr^+ ion implantation and pulsed ion-beam annealing on the formation and optical properties of Si/$CrSi_2$/Si(111) heterostructures. Technical Physics. 2010;**55**:1036-1044. DOI: 10.1134/S1063784210070194

[4] Умирзаков БЕ, Нормурадов МТ, Ташмухамедова ДА, Ташатов АК. Наноэпитаксиальные пленки и геттероструктуры на основе кремния. Ташкент: Merieus; 2012. 184 p. (in Russian)

[5] Normuradov MT, Khozhiev ST, Akhmedova LB, Kosimov IO, Davlatov MA, Dovranov KT. Peculiarities of BaTiO3 in electronic and X-ray analysis. In: E3S Web of Conferences. Vol. 383. 2023. p. 04068. DOI: 10.1051/e3sconf/202338304068

[6] Dovranov KT, Normurodov M, Davranov KhT, Bekpulatov IR. Formation of Mn4Si7/Si(111), CrSi2/Si(111), and CoSi2/Si(111) thin film and evaluation of their optically direct and indirect band gap. Ukrainian Journal of Physics. 2024;**20**(69). DOI: 10.15407/ujpe69.1.20

[7] Umirzakov BE, Bekpulatov IR, Turapov IK, Igamov BD. Effect of deposition of submonolayer Cs coatings on the density of electronic states and energy band parameters of $CoSi_2$/Si(111). Journal of Nano- and Electronic Physics. 2022, (4pp);**14**(2):02026. DOI: 10.21272/jnep.14(2).02026

[8] Badawi A, Althobaiti MG, Alharthi SS, Al-Baradi AM. Tailoring the optical properties of CdO nanostructures via barium doping for optical windows applications. Physics Letters A. 2021;**411**:127553. DOI: 10.1016/j.physleta.2021.127553

[9] Abdullahi SS, Güner S, Koseoglu Y, Musa IM, Adamu BI, Abdulhamid MI. Sımple method for the determination of band gap of a nanopowdered sample using Kubelka Munk theory. Journal of the Nigerian Association of Mathematical Physics. 2016;**35**:241-246

[10] Geiger R, Zabel T, Sigg HS. Group iv direct band gap photonics: Methods, challenges, and opportunities. Frontiers in Materials. 2015;**2**:52. DOI: 10.3389/fmats.2015.00052

[11] Dolgonosa A, Masona TO, Poeppelmeier KR. Direct optical band gap measurement in polycrystalline semiconductors: A critical look at the Tauc method. Journal of Solid State Chemistry. 2016;**240**:43-48. DOI: 10.1016/j.jssc.2016.05.010

[12] Latypov KF, Dolomatov MY. Estimation of the bandgap width of organic semiconductors photoconductivity by integral parameters of autocorrelational functions. Photonics. Optoelectronic Instruments & Devices. 2020;**14**(2):184-191. DOI: 10.22184/1993-7296.FRos.2020.14.2.184.191

[13] Normuradov MT, Khozhiev ST, Dovranov KT, Davranov KT,

Davlatov MA, Khollokov FK. Development of a technology for the production of nano-sized heterostructured films by ion-plasma deposition. Structure of materials. Ukrainian Journal of Physics. 2023;**68**(3):210. DOI: 10.15407/ujpe68.3.210

[14] Liu H, Wang L, Jiang Y, Li S, Liu D, Ji Y, et al. Study on SiO_2 thin films modified by post hot isostatic pressing. Vacuum. 2018;**148**:258-264. DOI: 10.1364/OIC.2016.MC.11

[15] Романов ИА, Ковальчук НС, Власукова ЛА, Пархоменко ИН, Солодуха ВА, Пилипенко ДВ, et al. Электролюминесценция пленок SiO_2 на Si, полученных термическим окислением и плазмохимическим осаждением. Журнал Белорусского государственного университета. Физика. 2021;**3**:26-31. DOI: 10.33581/2520-2243-2021-3-26-31

[16] Yeh C-F, Chen T-J, Fan C-L. Investigation of silicon oxide prepared by room-temperature ion plating. Journal of Applied Physics. 1998;**83**(2):1107-1113. DOI: 10.1063/1.366801

[17] Нормурадов МТ, Нормуродов ДА, Давронов ҚТ, Мустафаева НМ. Creation of new materials based on dielectric films using low-energy ion implantation. Euroasian Journal of Semiconductors. Science and Engineering. 2019;**1**(6):12-15, Article 8

[18] Bekpulatov IR, Imanova GT, Kamilov TS, Igamov BD, Turapov IKh. Formation of n-type CoSi monosilicide film which can be used in instrumentation. International Journal of Modern Physics B. 2022. 10 p. Article ID: 2350164. DOI: 10.1142/S0217979223501643

[19] Capan I, Pivac B, Slunjski R. Electrical characterisation of Si-SiO_2 structures. Physica Status Solidi C: Current Topics in Solid State Physics. 2011;**8**(3):816-818. DOI: 10.1002/pssc.201000076

[20] Некрашевич СС, Гриценко ВА. Электронная структура окисиа кремния. Физика твердого тела, том 56, вып. 2. 2014, С. 209-223

[21] Queeney KT, Weldon MK, Chang JP, Chabal YJ, Gurevich AB, Sapjeta J, et al. Infrared spectroscopic analysis of the Si/Sio_2 interface structure of thermally oxidized silicon. Journal of Applied Physics. 2000;**87**:1322-1330. DOI: 10.1063/1.372017

Chapter 4

New Analytical Potential, Charge, and Current Distributions in JL MOSFETs: Differences between Accurate and Simplified Models

Pedro Pereyra

Abstract

The operation of metal-oxide-semiconductor-field-effect transistors (MOSFETs) cannot be conceived without the prior formation of inversion layers. Charge, potential, and current distributions in this layer are important quantities for all types of MOSFET configurations. The mathematical difficulty to solve analytically the nonlinear Poisson equation in Kingston-Neustadter (KN) model, where the charge density includes electrons and holes, led to Hauser and Littlejohn (HL) to consider a simplified but solvable model, of only electrons or only holes, and to numerous HL-like compact analytical models. Recently, a new and simple method that overcomes the mathematical difficulty to solve the nonlinear Poisson equation in the inversion layer of a MOS has been introduced, and the more accurate Kingston-Neustadter model was successfully solved. We summarize here the new method and given the analytical solutions for the KN and HL models, we compare the potential and charge distribution predictions and show that they may differ by orders of magnitude. We also briefly outline the analytical results for the inversion layer width, the effective ionized atoms concentration, and the drift-diffusion currents in single and double gate junctionless (JL) MOSFETs. Specific calculations of these quantities, as functions of impurity concentration, gate potential, and oxide layer width, will be shown and compared with the predictions of the most exemplary simplified model, the Hauser-Littlejohn model.

Keywords: potential distribution in the inversion layer, the inversion-layer width, the electric potential in a MOS, effective impurity ionization, single and double gate JL MOSFET currents

1. Introduction

The calculation of the electric potential in the inversion layer of a MOS requires the solution of the Poisson equation, whose complexity depends on how the charge density is modeled. The charge density of the inversion layer emerges as soon as the gate potential V_G becomes larger than the threshold potential V_{Gt} and the conduction band edge crosses the Fermi level. Among the first attempts to calculate the electric potential in the inversion layer of a MOS, stand out the Shockley theoretical work on semiconductors physics [1] and the Kingston and Neustadter [2] approach of 1955, where a charge density of electrons and holes is assumed, in general. Kingston and Neustadter were able to analytically perform the first integration of the nonlinear Poisson equation, but not the second integration. Since then, the analytical efforts to obtain the potential distribution for this accurate model were virtually abandoned and alternative methods, related to the calculation of MOS charges, capacitance, and currents, were devised to account for the analysis of the physical and phenomenological properties of the MOS and the field effect transistors [3–7]. Some years later, in 1968, Hauser and Littlejohn [8] considered a simplified but solvable model in which the charge density contains only electrons or only holes. This is also the characteristic of the numerous analytical compact models introduced since then [9–23]. The literature on the operation of the MOS is overwhelming. A good reference for the standard approaches is Tisividis book [24].

Recently, the original problem of analytically solving the nonlinear Poisson equation in the inversion layer was successfully addressed [25]. In Section 2, we outline the inversion layer problem and the Kingston-Neustadter and Hauser-Littlejohn models. In Section 3, we introduce the bandstructure, relevant parameters, and basic relations to characterize a polarized metal oxide semiconductor. In Section 4, we present a summary of the new method that, based on displacement currents and fields arguments, made possible to replace the nonlinear Poisson equation of the inversion layer by a solvable second-order nonlinear differential equation, the solution of which solves also the original Poisson equation. In Section 5, we present the analytical formulas for the width of the inversion layer and the effective concentration of ionized impurities. In Section 6, we present the potential distributions for the KN and HL models, in a unified representation [25, 26]. We show that although these potentials are qualitatively similar, the predicted inversion layer widths are different. The same occurs with the charge distributions for the KN and HL models, studied in Section 7, as a function of the impurity concentration and the gate potential [26]. In Section 8, we present the analytical formula for the drift-diffusion current for a single-gate JL MOSFET and, in Section 9, the drift-diffusion current for a double-gate JL MOSFET [26].

2. The Kingston-Neustadter and the Hauser-Littlejohn models

In the original semiconductor theories of W. Shockley [1] it was assumed that in the most general case, there were electrons, holes as well as ionized donors and acceptors, and the charge density could be written as [1, 27]

$$\rho(y,z) = q\left(p(y,z) - n(y,z) - N_A^- + N_D^+\right), \tag{1}$$

where q is the electric charge, N_A^- and N_D^+ are the ionized acceptor and donor concentrations, and

$$n(y,z) = n_p e^{e\phi(y,z)/k_BT}, \text{ and } \quad p(y,z) = p_p e^{-e\phi(y,z)/k_BT}. \tag{2}$$

where k_B is the Boltzmann constant, T is the temperature, $\phi(y,z)$ is the local electric potential distribution of electrons and holes, and n_p and p_p are the minority and majority charge concentrations, which in terms of the intrinsic concentration n_i and of the Fermi and thermal potentials, ϕ_F and $\phi_T = k_BT/e$, can be written as

$$n_p = n_i e^{-\phi_F/\phi_T}, \text{ and } \quad p_p = n_i e^{\phi_F/\phi_T} \tag{3}$$

At high temperatures N_A^- and N_D^+ approximately equal to N_A and N_D, respectively. In a metal-oxide-semiconductor, the Kingston-Neustadter model assumption that the semiconductor has only one type of impurity, say type p, is accurate and there is no loss in generality when it is assumed that the charge of the ionized impurity concentration is N_A, which because of charge neutrality, deep in the bulk is $N_A = p_p - n_p$. Thus, in the KN model, the density can be written as [24]

$$\rho(y,z) = e\left[p_p e^{-\phi(y,z)/\phi_T} - n_p e^{\phi(y,z)/\phi_T} - p_p + n_p\right)\Big]. \tag{4}$$

Using the mass action law $n_p = n_i^2/p_p$, the relation $n_i/p_p = e^{-\phi_F/\phi_T}$ and the fact that $p_p \simeq N_A$, the charge density of electrons and holes for a type p semiconductor is written as [24]

$$\rho(z) = eN_A\left[e^{-\phi(y,z)/\phi_T} - 1 - e^{-2\phi_F/\phi_T}\left(e^{\phi(y,z)/\phi_T} - 1\right)\right]. \tag{5}$$

Before a source-drain potential is applied, the 1D approximation is good enough and the density, in the KN model, can be written as

$$\rho(z) = eN_A\left[e^{-\phi(z)/\phi_T} - 1 - e^{-2\phi_F/\phi_T}\left(e^{\phi(z)/\phi_T} - 1\right)\right]. \tag{6}$$

A common procedure to perform the first integration of the Poisson equation

$$\frac{d^2\phi}{dz^2} = -\frac{\rho(z)}{\varepsilon_s}, \tag{7}$$

is to use the identity

$$\frac{d}{dz}\left(\frac{d\phi}{dz}\right)^2 = 2\frac{d\phi}{dz}\frac{d^2\phi}{dz^2}, \tag{8}$$

and transform the Poisson equation into

$$\frac{d\phi}{dz} d\left(\frac{d\phi}{dz}\right) = -d\phi\frac{\rho(z)}{\varepsilon_s}. \tag{9}$$

A first integration of this equation can easily be performed. Kingston and Neustadter obtained the electric field in the inversion layer as

$$\begin{aligned}\frac{1}{2}\left(\frac{d\phi(z)}{dz}\right)^2 &= \frac{2eN_A}{\varepsilon_s} e^{-\phi_F/\phi_T}\left(\phi_T \cosh\frac{\phi(z) - \phi_F}{\phi_T} + \phi(z)\sinh\frac{\phi_F}{\phi_T} - \phi_T\cosh\frac{\phi_F}{\phi_T}\right)\\ &\equiv G(z, \phi(z)).\end{aligned} \tag{10}$$

Here $e\phi_F$ is the intrinsic Fermi energy. This first-order nonlinear differential equation could not be integrated once more. Since then, alternative, approximate, and numerical approaches have been introduced. Our purpose here is to show, in the next sections, that this problem has been overcome.

Some years later, in 1968, Hauser and Littlejohn considered a simplified but solvable charge density model where the difference p-n is replaced by only p or n, and the impurity concentration is N_A or N_D, depending on whether the space-charge region is p-type or n-type. If it is type p, the charge density in the inversion layer is written as

$$\rho_H(z) = -eN_A e^{\phi_H(z)/\phi_T}, \tag{11}$$

and the first integration, using also the identity (8), gives the electric field

$$\frac{1}{2}\left(\frac{d\phi_H(z)}{dz}\right)^2 = \frac{eN_A\phi_T}{\varepsilon_s}\left(e^{\phi_H(z)/\phi_T} - 1\right) + \frac{1}{2}\mathcal{E}_b^2, \tag{12}$$

with $\mathcal{E}_b$ the electric field at $z = z_b$, the boundary between the inversion and depletion layer, where the HL model assumes that the potential vanishes, that is, $\phi_H(z_b) = 0$. This first-order differential equation can be integrated and the electrostatic potential becomes [8].

$$\phi_H(z) = \phi_T \ln\left[-\frac{b}{a}\mathrm{sech}^2\left(\frac{(z - z_b)\sqrt{b}}{2\phi_T} + \tanh^{-1}\sqrt{\frac{a}{b} + 1}\right)\right], \tag{13}$$

with

$$a = \frac{2eN_0\phi_T}{\varepsilon_s} \quad \text{and} \quad b = \mathcal{E}_b^2 - a. \tag{14}$$

Notice that $\phi'_H(z_b) = 2qN_0 V_{Gu}/\varepsilon_s \equiv -\mathcal{E}_b$. Here $V_{Gu} = V_{Gt} + \mathcal{E}^t_{ox} t_{ox}$, with V_{Gt} the threshold gate potential, $\mathcal{E}^t_{ox}$ the electric field in the oxide layer at threshold, and t_{ox} the oxide layer width, see **Figure 1**. In the following, we will refer to these models as the KN and HL models and one of our purposes is to compare their predictions. It is worth mentioning that besides these analytical attempts, specific analytical solutions for double-gate MOSFETs, using basically the HL model, were obtained by, among others, Y. Taur and by A. Ortiz-Conde et al. We will not address these works nor the numerous analytical compact models, nor the numerical approaches and simulation codes.

In the last 20 years, new geometries and multi-gate MOSFETs were introduced. The inversion layer phenomenon that occurs in the oxide-semiconductor interface of MOS in a single-gate (SG) MOSFET occurs also in each oxide-semiconductor interface of multi-gate MOSFETs. Hence, studying the inversion layer in a single MOS structure is essential. After obtaining the potential and charge distributions, as explicit functions of the relevant parameters, we will consider the analytical expression for the current distribution in a single gate JL MOSFET, and based on this theory, we will discuss the calculation of the analytical formulas for the potential distribution and current in a double gate JL MOSFET, as a simple extension of the SG MOSFET theory.

3. Parameters and basic relations in biased MOS structures

To fix the boundary conditions, the notation and some of the essential parameters used to characterize the MOS structure, we recall here some well-established relations and definitions. Most of the of these relations can be found in textbooks, and some good references are [1, 6, 24, 27–31]. We will assume that the potentials are measured from the flat-band condition and that the electric potential at z_p (and beyond this point, i.e., in the bulk (see **Figure 1**)), is $\phi(z_p) = 0$. For gate potentials larger than the threshold potential, we will distinguish three characteristic layers

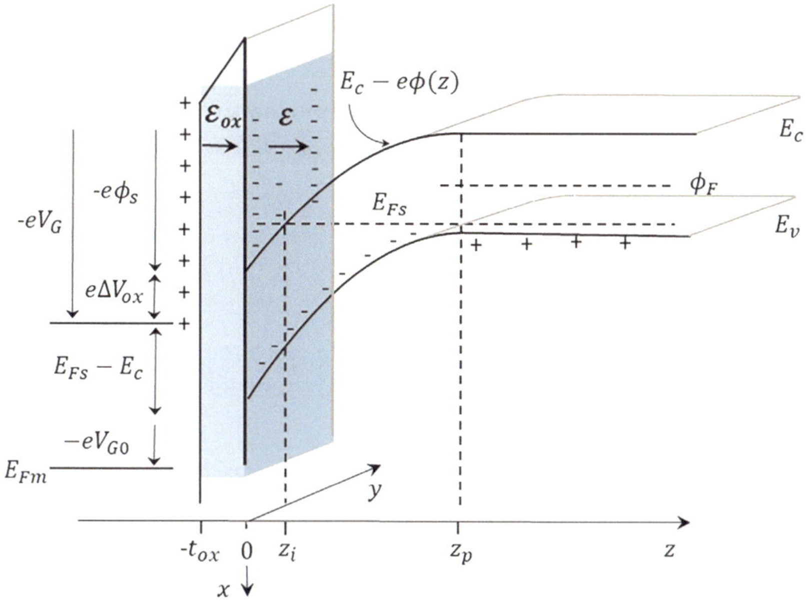

Figure 1.
A band structure of a metal-oxide-semiconductor with a type-p semiconductor and a positive gate potential V_G larger than the threshold potential. Several parameters that characterize the system are shown.

between the $z = -t_{ox}$ and z_p. Between $-t_{ox}$ and $z = 0$ the oxide-layer, where the electric potential is linear and given by:

$$\phi_{ox}(z) = V_G - (z + t_{ox})\mathcal{E}_{ox} \; for \quad -t_{ox} < z < 0, \tag{15}$$

and the capacitance per unit area is

$$C_{ox} = \frac{\varepsilon_{ox}}{t_{ox}}, \tag{16}$$

with ε_{ox} the electric permittivity in the oxide layer. From $z = 0$ to z_i we have the inversion layer, and from z_i to z_p, the depletion layer. We will assume that the charge density of the ionized impurity atoms in the depletion layer is homogeneous and will be denoted as N_d. We will see below that $N_d \leq N_A$, being N_A the concentration of acceptor impurities. The precise concentration of the ionized impurity atoms will be determined by the continuity requirements. A closed formula will be obtained for this dynamic quantity.

The boundary conditions at the oxide-semiconductor interface, that is, at $z = 0$, and at the edge of the depletion layer, say at z_p, are

$$\phi_{ox}(0) = V_G - \mathcal{E}_{ox} t_{ox}, \tag{17}$$

and

$$\phi_d(z_p) = 0, \quad d\phi_d(z_p)/dz = 0, \tag{18}$$

respectively. Under these conditions, the electric potential distribution in the depletion layer is given by

$$\phi_d(z) = \frac{eN_d}{2\varepsilon_s}\left(z_p - z\right)^2 + \phi\left(z_p\right) \quad z_i < z < z_p, \tag{19}$$

with $z_p = z_i + w_p$, and $w_p = \sqrt{2\varepsilon_s V_{Gu}/eN_A}$ the depletion layer width. The electric field, as a function of the electric potential, can be written as

$$\mathcal{E}_d(z, \phi_d(z)) = \sqrt{\frac{2eN_d\phi_d(z)}{\varepsilon_s}} \quad z_i < z < z_p. \tag{20}$$

Here and in the following, we write ε_s instead of $\varepsilon_r\varepsilon_0$.

As is well known, for the band-edge bending to reach the threshold potential, as shown in **Figure 1**, the gate potential has to be greater than the threshold potential V_{Gt}, defined by

$$V_{Gt} = \left(E_g - E_{Fs} + \mathcal{E}_{ox}^t t_{ox}\right)/e. \tag{21}$$

Here, E_g is the gap energy, E_{Fs} the semiconductor Fermi energy, $\mathcal{E}_{ox}^t$ the electric field in the oxide layer at threshold, and t_{ox} the oxide layer width. Once the inversion regime is established and V_G is increased, the highly charged inversion layer emerges whose thickness, the inversion-layer width z_i, depends on the gate potential and on the impurity concentration. In terms of the parameters defined above, the condition for determining the inversion layer width is

$$V_{Gt} = \mathcal{E}_{ox}^t t_{ox} + \phi(z_i), \tag{22}$$

where

$$\phi(z_i) = \frac{E_c - E_{Fs}}{e} \equiv V_{Gu}, \tag{23}$$

is the surface potential threshold denoted here as V_{Gu}, with u for *umbral* (from latin *umbra*—shadow, and *liminaris*—limit). After solving Poisson's equation, we will present a closed formula for the inversion-layer width z_i.

The positive charge on the metallic side creates an electric field $\mathcal{E}(z)$ that penetrates on the semiconductor side, attracts electrons, and repels holes and, at the end, establishes an electric potential $\phi(z)$ that redefines the valence and conduction band edges as $E_v - e\phi(z)$ and $E_c - e\phi(z)$, respectively. The inversion population process modifies the nature and the phenomenology of the physical system on the semiconductor side, whose precise description depends on whether the Poisson equation that governs the intertwined relation, between gate potential, charge distribution, and electric potential, can be fully or partially solved. In the next section, we summarize the main argument to transform the Poisson equation into a solvable nonlinear second-order differential equation.

4. Charge and field displacement in the inversion layer

When the gate potential V_G becomes greater than the threshold potential V_{Gt}, the charge distribution ρ_i, regardless of how it is modeled, undergoes an inversion-population process (IPP). If the semiconductor is linear and the charge concentration $\mathcal{N}_i(z, \phi, T)$ in the growing inversion layer is highly localized, the balance between the time rate of the thermal energy $k_B T\mathcal{N}_i(z,\phi,T)$ and the time rate of the displacement field energy $u_E(\phi,T) = t\cdot\mathcal{D}$, can be written as

$$-\phi_T\int_{\rho_i(\phi_t,T)}^{\rho_i(\phi,T)} d\rho_i(\phi,T) = \frac{1}{2}\int_{u_E(\phi_t,T)}^{u_E(\phi,T)} du_E. \tag{24}$$

where $\rho_i(\phi,T) = -e\mathcal{N}_i(z,\phi,T)$ is the charge distribution in the inversion layer, and ϕ_t the electric potential at threshold. Assuming that $\mathcal{E} = -d\phi(z)/dz\hat{\mathbf{z}}$, we have

$$\frac{\varepsilon_s}{2}\left(\frac{d\phi(z)}{dz}\right)^2 - \frac{\varepsilon_s}{2}\mathcal{E}_t^2 = -\phi_T(\rho_i(\phi) - \rho_i(\phi_t)). \tag{25}$$

It was shown that when the charge density is, for example, the charge density of the KN model, the electric field obtained from this equation is equal to the electric field in Eq. (10). Taking into account the Poisson equation and the electric field in Eq. (10), we obtain the nonlinear second-order differential equation

$$\phi_T\frac{d^2\phi(z)}{dz^2} = \frac{1}{2}\left(\frac{d\phi(z)}{dz}\right)^2 - \frac{1}{2}\xi^2, \tag{26}$$

where ξ is a model and threshold-dependent parameter, given by

$$\xi^2 = \mathcal{E}_t^2 + 2\frac{\phi_T\rho(\phi_t)}{\varepsilon_s}. \tag{27}$$

These are the main and most consequential results of the new method. In the next sections, we will apply this method for the HL and the KN models. The charge densities in these models are different; hence, the parameter ξ is also different, while the differential equations are formally similar. Eq. (26) shows an important step toward the solution of the Poisson equation in the inversion layer. Eq. (26) is solvable and, as will be seen below, its solution is also a solution of the original Poisson equation. It has been shown in Ref. [25], that the analytical solution of (26) is

$$\phi(z) = \phi_0 - 2\phi_T\ln\left(\cosh\frac{\xi(z-z_0)}{2\phi_T} - \frac{f_0}{\xi}\sinh\frac{\xi(z-z_0)}{2\phi_T}\right), \tag{28}$$

with ϕ_0 the potential and f_0 the electric field at $z = z_0$. Given this solution, and before we refer to the electric potential in the HL and KN models, it is convenient to consider the inversion layer width in general.

5. Inversion layer width and depletion layer concentration

When the continuity conditions are imposed at z_i, one obtains both the inversion layer width

$$z_i = \frac{2\phi_T}{\xi} \cosh^{-1}\left(\frac{-e^{\theta/2\phi_T}\xi^2 + \sqrt{\mathcal{E}_s^4 - \mathcal{E}_s^2\xi^2(1 - e^{\theta/\phi_t})}}{\mathcal{E}_s^2 - \xi^2} \right), \tag{29}$$

where $\theta = \phi_s - V_{Gu}$, and the concentration of the ionized impurity atoms in the inversion layer regime, given by

$$N_{dl} = \frac{\varepsilon_s}{2e\phi(z_i)} \left(\xi^2 + e^{-(\phi_s - \phi(z_i))/\phi_t} \left(\mathcal{E}_s^2 - \xi^2 \right) \right), \tag{30}$$

when $\theta \geq 0$, and by

$$N_{ds} = \frac{\varepsilon_s \mathcal{E}_s^2}{2e\phi_s}, \tag{31}$$

when $\phi_s < \phi(z_i)$. Replacing the corresponding parameter ξ one has the inversion layer width predicted in the corresponding model. See the next section.

6. Potential distributions

As mentioned above, the difference between one model and another lies in the charge density and therefore in the parameter ξ. Now we will be more specific.

6.1 Potential distribution in Hauser-Littlejohn model

In the solvable Hauser-Littlejohn model, assuming a type p semiconductor, the charge density and the parameter ξ are

$$\rho_H(z) = -eN_0 e^{e\phi_H(z)/k_B T} \quad \text{and} \quad \xi_H^2 = \mathcal{E}_b^2 - \frac{2eN_0\phi_T}{\varepsilon_s}. \tag{32}$$

For the boundary conditions considered by Hauser and Littlejohn, that is, for $z_0 = z_b$, and at this point $\phi_0 = \phi_b = 0$ and $f_0 = -\mathcal{E}_b$, the potential distribution becomes

$$\phi_H(z) = -2\phi_T \ln\left(\cosh \frac{\xi_H(z - z_b)}{2\phi_T} + \frac{\mathcal{E}_b}{\xi_H} \sinh \frac{\xi_H(z - z_b)}{2\phi_T} \right), \tag{33}$$

with $\mathcal{E}_b = \sqrt{2eN_d V_{Gu}/\varepsilon_s}$. It is not difficult to show that the HL solution in (13), obtained by direct integration, reduces *rigorously* to $\phi_H(z)$ (see also Appendix of Ref. [25]). In this model, $z_b = z_i$ defines the boundary between the inversion and depletion layer, and Hauser and Littlejohn assume, that the electric potential at this point vanishes, and we can write also the potential distribution for the HL model in terms of the surface parameters at the oxide-semiconductor interface, that is, at $z_0 = 0$, where $\phi_0 = \phi_s$ and $f_0 = -\mathcal{E}_{Hs}$. In this case, the potential distribution is written as

$$\phi_L(z) = \phi_s - 2\phi_T \ln\left(\cosh \frac{\xi_H z}{2\phi_T} + \frac{\mathcal{E}_{Hs}}{\xi_H} \sinh \frac{\xi_H z}{2\phi_T} \right). \tag{34}$$

Here the surface electric field $\mathcal{E}_{Hs}$ is

$$\mathcal{E}_{Hs}(z) = \left(-eN_A\phi_T e^{e(\phi_s - V_{Gu})/\phi_T} + \xi_H^2\right)^{1/2}. \tag{35}$$

Notice that the assumption $\phi_H(z_b) = 0$, implies that $\phi_H(z) = \phi_L(z) - V_{Gu}$. In terms of $\phi_L(z)$, the charge density in the Hauser-Littlejohn model is

$$\rho_H(z) = -eN_A e^{e(\phi_L(z) - V_{Gu})/k_B T}. \tag{36}$$

In **Figure 2** we plot the potential energy $-e\phi_L(z)$ as function of z, for different values of the impurity concentration, and for different surface potentials, as well. In the next subsection, by comparing with the more accurate KN model, it will be seen that the HL model predicts much smaller inversion layer widths, and grows faster near the oxide-semiconductor interface. The potential distribution $\phi_H(z)$ is well known and has been widely used as an approximate result for double-gate devices [15, 16].

6.2 Potential distribution in the Kingston-Neustadter model

In the important Kingston-Neustadter model, the charge density can be written as

$$\rho_K(z) = 2eN_A e^{-\phi_F/\phi_T}\left[\sinh\frac{\phi_F - \phi_K(z)}{\phi_T} - \sinh\frac{\phi_F}{\phi_T}\right]. \tag{37}$$

The parameter ξ is

$$\xi_K^2 = \frac{8eN_A}{\varepsilon_s} e^{-\phi_F/\phi_T}(\phi_F - \phi_T)\sinh\frac{\phi_F}{\phi_T}, \tag{38}$$

and the potential distribution, for $\phi_0 = \phi_s, f_0 = -\mathcal{E}_s$ and $z_0 = 0$, becomes

$$\phi_K(z) = \phi_s - 2\phi_T \ln\left(\cosh\frac{\xi_K z}{2\phi_T} + \frac{\mathcal{E}_s}{\xi_K}\sinh\frac{\xi_K z}{2\phi_T}\right). \tag{39}$$

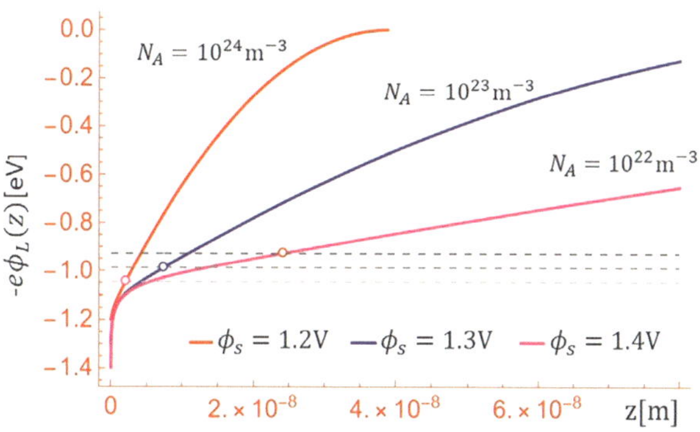

Figure 2.
The potential energy in the HL model as a function of z for different values of the impurity concentration N_A and different surface potentials ϕ_s. The small circles indicate the limit (at z_i) between the inversion and the depletion layers. The dashed black lines define the threshold energies $-eV_{Gu}$, in eVs.

Given this potential distribution, an alternative representation for the electric field in the inversion layer is

$$\mathcal{E}_K(z) = \xi_K \frac{\sinh \frac{\xi_K z}{2\phi_T} + \frac{\mathcal{E}_s}{\xi_K} \cosh \frac{\xi_K z}{2\phi_T}}{\cosh \frac{\xi_K z}{2\phi_T} + \frac{\mathcal{E}_s}{\xi_K} \sinh \frac{\xi_K z}{2\phi_T}}. \tag{40}$$

Here the surface electric field is

$$\mathcal{E}_s = \left[\frac{2eN_A}{\varepsilon_s} e^{-\phi_F/\phi_T} \left(\phi_T \cosh \frac{\phi_s - \phi_F}{\phi_T} + \phi_s \sinh \frac{\phi_F}{\phi_T} - \phi_T \cosh \frac{\phi_F}{\phi_T}\right)\right]^{1/2}. \tag{41}$$

It was shown in Ref. [25]), by substitution and numerical evaluation, that the function $\phi_K(z)$ in (39) solves also the original Poisson equation and the electric field in (40), matches with the electric field $\sqrt{2G(z, \phi(z))}$ in (10), that was obtained by Kingston and Neustadter after the first integration of the original Poisson equation J. Reiter has shown also [32], by numerically solving the differential equation, that the difference between the numerical and analytical solutions is less than 0.5%, in the entire parametric space.

In **Figure 3** we plot the potential energy $-e\phi_K(z)$ as functions of z, for different values of the impurity concentrations N_A, as well as for different values of the surface potential ϕ_s. In this figure, we indicate the Fermi levels (dashed black lines) and the borders between the inversion and depletion layers. These points, where the potential energies cross the Fermi levels, define the inversion layer widths z_i. The behavior of $\phi_K(z)$ is qualitatively similar to that of $\phi_L(z)$ in **Figure 2**. It grows also rapidly for gate potentials V_G larger than the threshold potential V_{Gt}, or surface potentials ψ_s larger than the umbral potential V_{Gu}.

It is now possible to compare the electric potentials and the inversion layer widths predicted by the Kingston-Neustadter and the Hauser-Littlejohn models. In **Figure 4,** we plot the potential energies, predicted in both models, as functions of z, and for different values of the impurity concentration N_A and different values of the surface potential ϕ_s. The potential distributions are qualitatively similar but quantitatively different. In the HL model, the inversion layer widths are smaller than those predicted in the KN model, by almost a factor of 2, the electric potentials are also smaller but

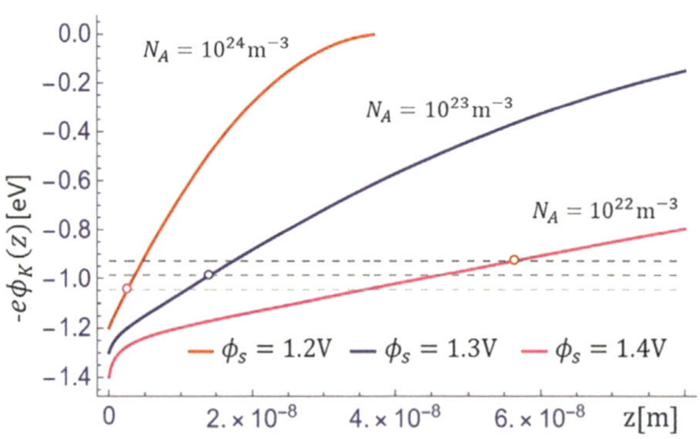

Figure 3.
The potential energy in the KN model as a function of z for different values of the impurity concentration N_A and different surface potentials ϕ_s. The small circles indicate the limit (at z_i) between the inversion and the depletion layers. The dashed black lines define the threshold energies $-eV_{Gu}$ (or Fermi levels).

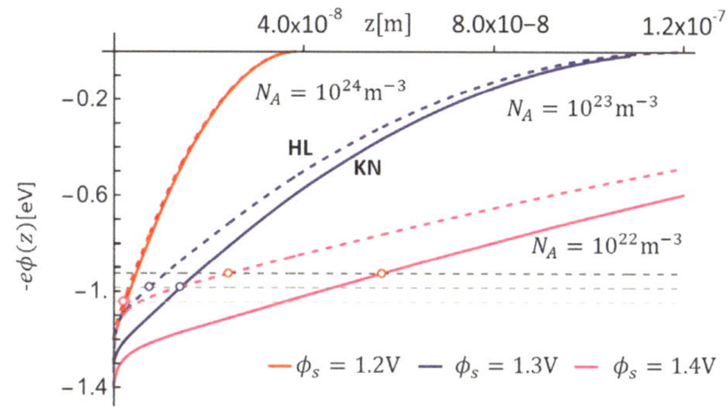

Figure 4.
The electric potential energy in the KN *(continuous) and* HL *(dashed) model as a function of z for different values of the impurity concentration N_A and different surface potentials ϕ_s. The small circles indicate the limit (at z_i) between the inversion and the depletion layers.*

near the oxide-semiconductor interface, the electric potentials in the HL model grow faster than in the KN model. In general, the potential and the layer widths are underestimated in the HL model. Because of the smaller inversion layer width and the rapid grow of $\phi_H(z)$, the charge distribution in the HL model is much more localized and closer to the 2D assumption.

In **Figure 5** we plot the inversion layer widths in the KN model as functions of the gate potential $V_G = \phi_s + \kappa t_{ox}\mathcal{E}_s$, for different values of the impurity concentration. For these plots, the silicon parameter $\kappa = \varepsilon_s/\varepsilon_{ox} = 3$ and oxide-layer width $t_{ox} = 2\text{nm}$ were considered.

Given the explicit functions ϕ_L, ϕ_K, ξ_H and ξ_K, we will in the next section evaluate the charge density distributions.

7. Charge density distributions

An important consequence of the inversion population phenomenon is the rapid increase of the charge density in the inversion layer. Replacing the electric potentials

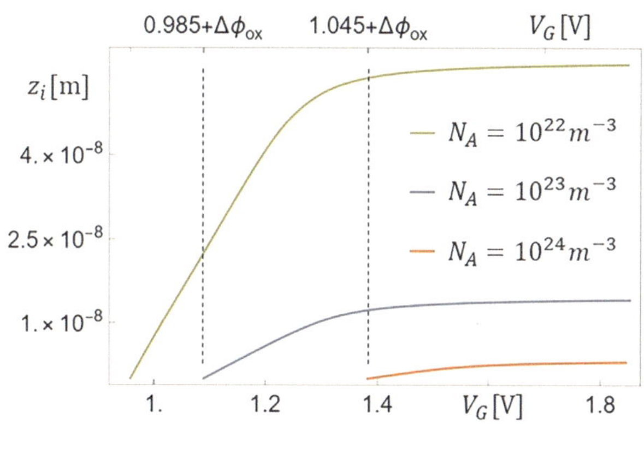

Figure 5.
The inversion-layer width z_i in the KN model as function of V_G, for different impurity concentrations. In the upper axes, the threshold potentials is written as $V_{Gt} = V_{Gu} + \Delta_{ox}$, with $\Delta_{ox} = t_{ox}E^t_{ox} = \kappa t_{ox}E^t_s$, the potential drop in the oxide layer.

$\phi_L(z)$ and $\phi_K(z)$ in the corresponding density, Eqs. (36) and (37) we have, for the HL model, the charge density

$$\rho_H(z) = -eN_{Hd}e^{(\phi_s - V_{Gu})/\phi_T}\left(\cosh\frac{\xi_H z}{2\phi_T} + \frac{\mathcal{E}_{Hs}}{\xi_H}\sinh\frac{\xi_H z}{2\phi_T}\right)^2, \quad (42)$$

and, for the Kingston-Littlejohn model, the charge density

$$\begin{aligned}\rho_K(z) = eN_{Kd}\Bigg\{ & e^{-\phi_s/\phi_T}\left(\cosh\frac{\xi_H z}{2\phi_T} + \frac{\mathcal{E}_s}{\xi_H}\sinh\frac{\xi_H z}{2\phi_T}\right)^2 - 1 \\ & - e^{-2\phi_F/\phi_T}\left[1 - e^{\phi_s/\phi_T}\left(\cosh\frac{\xi_H z}{2\phi_T} + \frac{\mathcal{E}_s}{\xi_H}\sinh\frac{\xi_H z}{2\phi_T}\right)^{-2}\right]\Bigg\}.\end{aligned} \quad (43)$$

To visualize the similarities and differences between these charge densities, we plot them side by side in **Figure 6**, and the surface densities in **Figure 7**. The densities in **Figure 6** are plotted as functions of the distance z to the oxide-semiconductor interface, for different values of the impurity concentration and for different values of the surface potential ϕ_s. The qualitative behavior is similar, with rapid growth in the inversion layer and almost constant densities in the depletion layers. In these layer, we take into account the effective impurity ionization N_d. The effect of this concentration is observed in **Figure 6a** where the charge density varies continuously at z_i and vanishes at the border of the depletion layer at $z_p = z_i + w_p$. The densities in the HL model behave as shown in **Figure 6b** with a practically constant concentration in the depletion layers. As observed before, the HL predicts thinner inversion layers. Near the inversion layer edge, around z_i, the densities are quantitatively similar, but as one approaches the oxide interface, the differences grow. To visualize this difference we plot in **Figure 7**, the surface densities

$$\rho_{Hs}(z) = -eN_{Hd}e^{(\phi_s - V_{Gu})/\phi_T}, \quad (44)$$

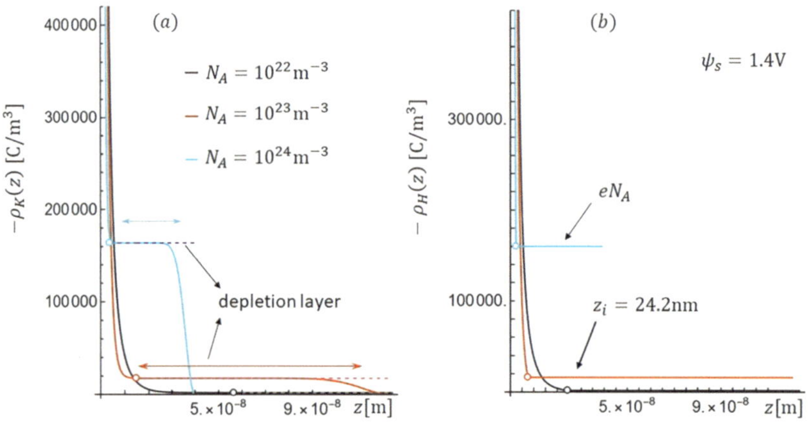

Figure 6.
Charge density in the KN *(left) and* HL *(right) models as a function of z, and for different values of* N_A. *The small circles indicate the limit (at* z_i*) between the inversion and the depletion layers.*

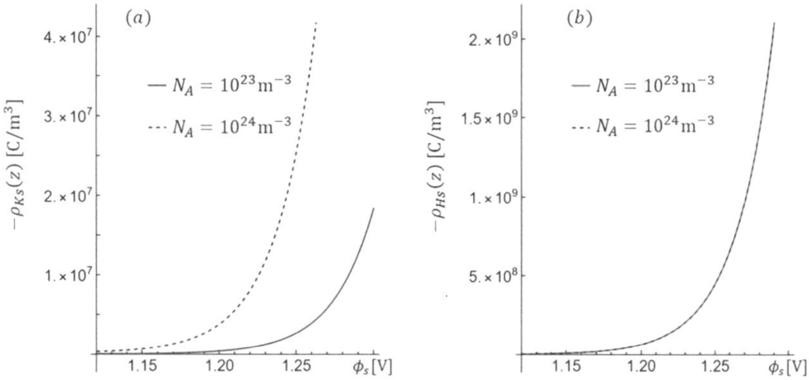

Figure 7.
The charge density at $z = 0$ in the KN (left) and the HL (right) model, as functions of the surface potential ϕ_s, for two values of the impurity concentration $N_A = 10^{23}m^{-3}$ and $N_A = 10^{24}m^{-3}$.

and

$$\rho_{Ks}(z) = -2eN_{Kd}e^{-\phi_F/\phi_T}\left(\sinh\frac{\phi_s - \phi_F}{\phi_T} + \sinh\frac{\phi_F}{\phi_T}\right) \tag{45}$$

From **Figure 7**, it is clear that the difference between the charge densities in the HL and KN models grows with ϕ_s by orders of magnitude. We also see that due to the exponential function in (44), the density at the surface, in the HL model, is practically insensitive to the impurity concentration, while in the KN model, the surface density in (45) is sensitive to the impurity concentration, as could be expected.

8. Drift current in a JL MOSFET channel

When the biased MOS system is part of a single or multiple gate MOSFETs, the charge and potential distributions obtained before play an important role in the nature and magnitude of the drift-diffusion currents moving in the MOSFET channel(s), parallel to the oxide-semiconductor interface(s). To simplify the analysis, we will have the so-called junctionless MOSFETs. As is well known and will be seen here, the charge carriers in the inversion layer play the most important role.

A well-established formula for the calculation of the source-drain current is Pao-Sah's formula [5]

$$I_D = \frac{eD_nW_x}{L}\int_0^{z_p} n(y,z)dz\int_0^L dy\frac{d\eta(y)}{dy}. \tag{46}$$

Here L is the channel length that we will assume alongside the y direction, W_x the channel depth, D_n the diffusion coefficient, and $z_p = z_i + w_p$ the inversion plus depletion layers widths. The semiconductor layer width W_z may be greater or smaller than z_p. When the drift and diffusion currents are taken into account, the distribution $n(y,z)$ becomes

$$n(y,z) = n_Ae^{(\phi(z)-\eta(y)-\phi_{Fs})/\phi_T}, \tag{47}$$

where $\phi(z)$ is the electrical potential due to the gate potential, $\eta(y)$ is the quasi-Fermi level modified by the drain-source potential U_D along the channel, and $\phi_{Fs} = (E_c - E_{Fs})/e$ is the equilibrium Fermi potential in the bulk, that we will assume to be measured from the conduction band edge. Usually dz is replaced by $-d\phi/\mathcal{E}(z,\phi)$, where $\mathcal{E}(z,\phi) = \sqrt{2G(z,\phi)}$. We do not need to make this change because we do have the electric potential as an explicit function of Z. We keep the integration upon z, but we will change the integration upon y by the integration upon η. We also split the integration in z in two parts, one for the inversion region and one for the depletion layer. In this case,

$$I_D = \frac{eD_n w_x n_i}{L}\left[\int_0^{z_i} e^{\phi_i(z)/\phi_T} dz + \int_{z_i}^{z_i+w_p} e^{\phi_d(z)/\phi_T} dz\right]\int_0^{U_D} e^{(-\eta(y)-\phi_{Fs})/\phi_T} d\eta, \tag{48}$$

with the function ϕ_i given by the electric potential defined in (39), and the function ϕ_d given above for the electric potentials in the depletion layer, Eqs. (30) and (31). Similarly, the current in the HL model is straightforwardly calculated when the electric potential defined in (34) is replaced for ϕ_i. In this section we will present results only for the KN model and we will drop the subscript K. The integrations are direct and we obtain, in the KN model, the current

$$\begin{aligned} I_D = \frac{eD_n w_x n_i}{L} & \left[e^{(\phi_s-\phi_{Fs})/\phi_T} \frac{2\phi_T \sinh\frac{\eta z}{2\phi_T}}{\eta\left(\cosh\frac{\eta z}{2\phi_T} + \frac{\mathcal{E}_s}{\eta}\sinh\frac{\eta z}{2\phi_T}\right)}\right|_0^{z_i} \\ & + \frac{\sqrt{\pi\varepsilon_s\phi_T}}{\sqrt{eN_d}} e^{-\phi_F/2\phi_T} \mathrm{Erf}\left[\frac{\sqrt{eN_d}w_p}{2\sqrt{\varepsilon_s\phi_T}}\right]\Bigg]\left(1 - e^{-U_D/\phi_T}\right), \end{aligned} \tag{49}$$

which reduces to

$$\begin{aligned} I_D = \frac{eD_n w_x n_i}{L} & \left[\frac{2\phi_T}{\xi} e^{(\phi_s-\phi_{Fs})/\phi_T} e^{(V_{Gu}-\phi_s)/2\phi_T} \sinh\frac{\xi z_i}{2\phi_T}\right. \\ & \left. + \frac{\sqrt{\pi\varepsilon_s\phi_T}}{\sqrt{eN_d}} e^{-\phi_F/2\phi_T} \mathrm{Erf}\left[\frac{\sqrt{eN_d}w_p}{2\sqrt{\varepsilon_s\phi_T}}\right]\right]\left(1 - e^{-U_D/\phi_T}\right). \end{aligned} \tag{50}$$

Since $V_{Gu} = \phi_{Fs}$, this becomes

$$\begin{aligned} I_D = \frac{eD_n w_x n_i}{L} & \left[\frac{2\phi_T}{\xi} e^{(\phi_s-V_{Gu})/2\phi_T} \sinh\frac{\xi z_i}{2\phi_T}\right. \\ & \left. + \frac{\sqrt{\pi\varepsilon_s\phi_T}}{\sqrt{eN_d}} e^{-\phi_F/2\phi_T} \mathrm{Erf}\left[\frac{\sqrt{eN_d}w_p}{2\sqrt{\varepsilon_s\phi_T}}\right]\right]\left(1 - e^{-U_D/\phi_T}\right) \equiv I_{Di} + I_{Dd}. \end{aligned} \tag{51}$$

As mentioned before, the concentration N_d should be replaced by N_{ds}, when $\phi_s < V_{Gu}$, and by N_{dl} when $\phi_s > V_{Gu}$. We shall present now some graphs to visualize the behavior of these currents when both the gate and source-drain potentials are applied. For all graphs shown in this section, we will assume the mobility $\mu = 400\mathrm{cm}^2/\mathrm{Vs}$ in the depletion and inversion layers. These quantities can of course be changed. The impurity concentrations N_A will be indicated in the graphs.

We will plot now the depletion and inversion layer currents and the total current, separately, as functions of the source-drain potential U_D and of the surface potential ϕ_s. Since $V_G = \phi_s + \kappa t_{ox}\mathcal{E}_s$, plotting as a function of ϕ_s is almost as plotting as function of V_G, not exactly as will be seen below. In all cases, of this section, we will assume that the mobility $\mu = 400\text{cm}^2/\text{Vm}$ is the same in the depletion and inversion layers, $L = 0.4\mu\text{m}$ and $W_x = 1.0\mu\text{m}$. In **Figure 8** we plot the depletion, inversion, and total currents as functions of U_D and of ϕ_s. In (a) the depletion current below and above the threshold, in (b) the inversión layer current and in (c) the sum of both currents. Above the threshold, the current I_{Dd} grows with U_D but is independent of V_G. Both currents I_{Dd} and I_{Di} grow monotonously with U_D and tend to a saturation value at higher values. For this graph, we assumed also that $W_x = 1\mu\text{m}$ and $L = W_x/2.5$.

Due to the nonlinear relationship between the gate and the surface potential and the nonlinear relationship between the surface electric field $\mathcal{E}_s$ and ϕ_s, the growths of the gate current inversion layer as a function of ϕ_s and as a function of V_G should look different. In fact, this difference can be seen in **Figure 9** (a) and (b), where the same inversion current is plotted, in one case as a function of ϕ_s, and as a function of V_G in the other case. In **Figure 9** (a) the current I_{Di} grows exponentially as a function of ϕ_s while the currents I_{Di} in **Figure 9** (b) grow almost linearly as function of V_G. Since the depletion current remains almost constant above the threshold behavior the linear behavior of I_{Di} occurs also in the total current I_D.

In **Figure 10**, we plot the total drift-diffusion current I_D as a function of the gate potential V_G, for different values of the impurity concentration. As mentioned before

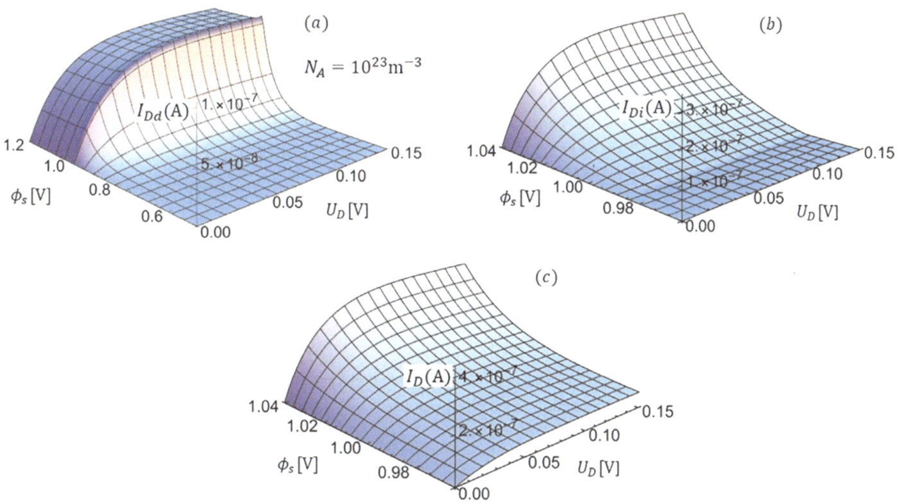

Figure 8.
Depletion, inversion, and total current as functions of the source-drain potential U_D and of the surface potential ϕ_s. In (a) the depletion current I_{Dd} below and above the threshold, in (b) the inversión layer current I_{Di} and in (c) the total current $I_D = I_{Dd} + I_{Di}$. The depletion current I_{Dd} as a function of U_D grow monotonously toward a saturation value, as a function of ϕ_s the depletion current grows steadily until the gate potential reaches the threshold value, where I_{Dd} experiences a sharp change. Above this threshold, the depletion current becomes independent of the gate potential. At variance with I_{Dd}, the inversión layer current I_{Di}, as a function of the ϕ_s (i.e. of the gate potential), starts growing at the threshold potential, and for a fixed value of the gate potential, the inversion layer current also grows with U_D and tends to a saturation value at higher values. For this graph $\mu = 400cm^2/Vs$, $W_x = 1\mu m$, $L = W_x/2.5$.

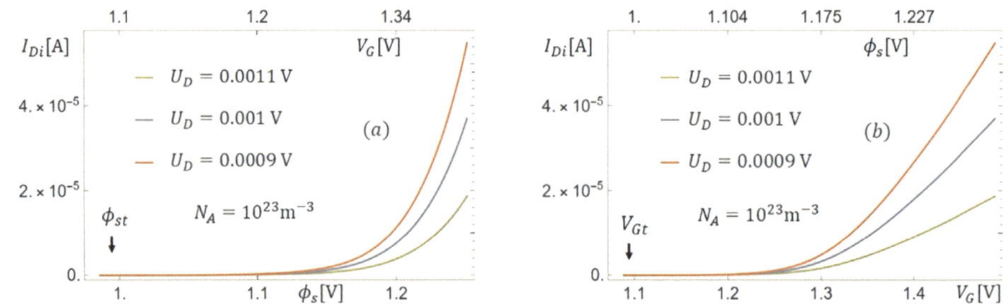

Figure 9.
The inversion-layer current I_{Di} as function of the surface potential ϕ_s (panel (a)) and the gate potential V_G (panel (b)). In (a) the current is plotted as a function of (equidistant values) of ϕ_s, and in (b), the same current but as a function of (equidistant values) of V_G. In the last case, it is evident that at larger values of V_G the current becomes a linear function of V_G.

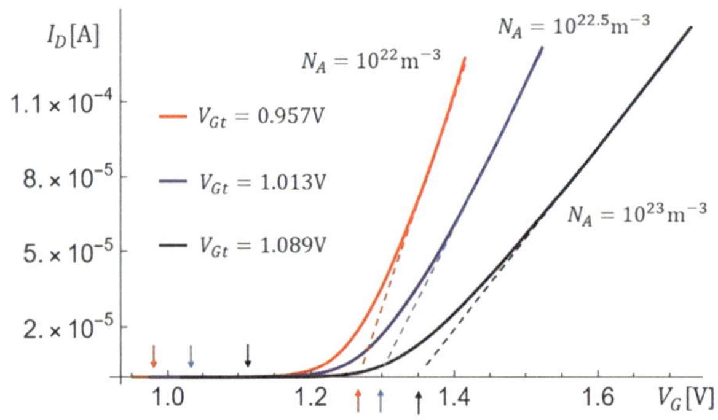

Figure 10.
The total source-drain current $I_D = I_{Dd} + I_{Di}$ in silicon MOSFETs as functions of the gate potential, for three different values of the impurity concentration. Here, we assume that the mobility, in the depletion and inversion layers, is $\mu = 400 cm^2/Vm$, $U_D = 0.001 V$, and channel length $L = 0.5 \mu m$. The gate potential thresholds V_{Gt} are given on the graph. These thresholds and the so-called operational threshold potentials, defined generally at the intersection of the abscissa with the extension of the linear current (see dashed lines), are indicated with arrows.

the inversion layer current and the total current behave, at larger values of the gate potential, as linear functions of this variable, with lower slopes at higher impurity concentrations. It can be seen also in **Figure 10**, that the threshold potential for higher impurity concentration is larger. These threshold potentials and the so-called *operational threshold potentials*, determined at the intersection of the linear current with the abscissa, are indicated with arrows. The operational threshold potentials are larger than the gate potential thresholds. In the particular case considered here, the differences are of the order of 0.3 V. Given the device parameters, the operational potentials can also be predicted. To complete the comparison of the KN and HL model predictions, we plot in **Figure 11** the drift-diffusion currents in the KN and HL models assuming the same device parameters. The predict currents are also different. The currents in the simplified HL model are smaller than the currents in the most accurate KN model. For these graphs, we assumed that the impurity concentration is $N_A = 10^{23} m^{-3}$.

The theory presented in this section, and before, can be applied almost straightforwardly to more complex devices, among them, to the double-gate (DG) MOSFET studied in the next section, and to multi-gate MOSFETs.

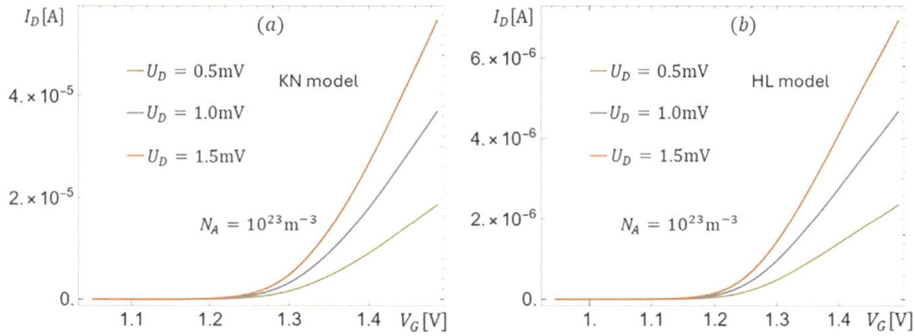

Figure 11.
The drift-diffusion currents in the KN *and* HL *models as functions of the gate potential. The* HL *model predicts smaller currents than the* KN *model.*

9. The double gate JL MOSFET

Explicit calculations for other MOSFET configurations can be performed almost as a trivial extension of the results presented in the last sections. We will refer only and briefly to the double-gate MOSFET. Consider a double-gate device like the one sketched in **Figure 12**, where the oxide layers are at a distance t_{Si} and the front and back gate voltages V_{Gf} and V_{Gb} are assumed to have opposite polarizations. In this case, holes accumulate near the back oxide-semiconductor interface and electrons near the front oxide-semiconductor. Before the source-drain potential U_D is applied, we can write the electric potential in the inversion layers in the 1D approximation as

$$\phi_{iDG}(z) = \phi_{if}(z) + \phi_{ib}(z). \tag{52}$$

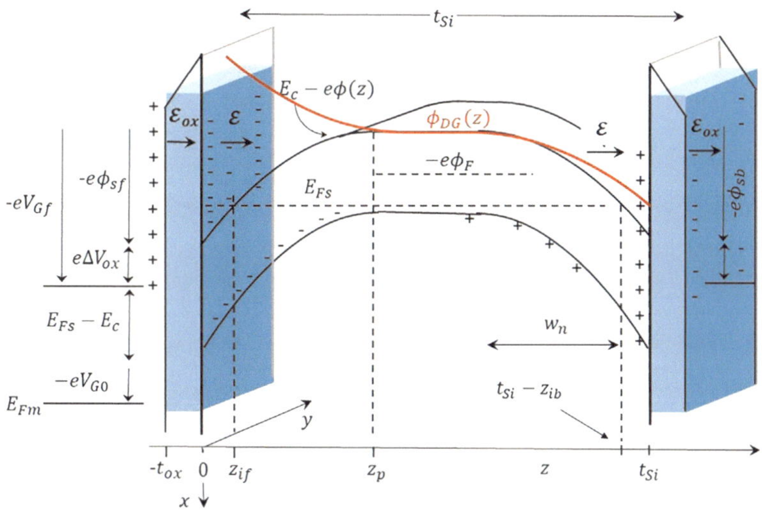

Figure 12.
A double gate structure as a generalization of the single gate MOS of **Figure 1**. *We assume here that the back gate potential is negative and attracts holes. The potential distribution $\phi_{DG}(z)$ (red curve) is schematically shown.*

Assuming that the charge densities are described by the Kingston-Neustadter-like distributions, the potential distributions in the front and back inversion layers, respectively, can be written as

$$\phi_{if}(z) = \phi_{sf} - 2\phi_T \ln\left(\cosh \frac{\xi_K z}{2\phi_T} + \frac{\mathcal{E}_{Kfs}}{\xi_K} \sinh \frac{\xi_K z}{2\phi_T} \right), \tag{53}$$

and

$$\phi_{ib}(z) = -\phi_{sb} + 2\phi_T \ln\left(\cosh \frac{\xi_K z'}{2\phi_T} + \frac{\mathcal{E}_{Kbs}}{\xi_K} \sinh \frac{\xi_K z'}{2\phi_T} \right). \tag{54}$$

Here, $z' = t_{Si} - z$, ϕ_{sf} and ϕ_{sb} are the front and back surface potentials, and $\mathcal{E}_{Kfs}$ and $\mathcal{E}_{Kbs}$ are the front and back surface electric fields. In the symmetric case with $t_{oxf} = t_{osb}$ and $V_{Gf} = -V_{Gb}$, the potential function $\phi_{DG}(z)$ is asymmetric, as sketched in **Figure 12** (red curve), and the potential energy $q\phi_{iDG}(z)$ is symmetric, as shown in **Figure 13** (*a*), where a silicon layer $t_{Si} \simeq 40\,\text{nm}$ is assumed, which for an impurity concentration $N_A = 1 \times 10^{18}\,\text{cm}^{-3}$ is $\simeq (1/2)(w_{pd} + w_{nd} + z_{if} + z_{ib})$, where w_{pd} and w_{nd} are the depletion layers widths and z_{if} and z_{ib} the inversion layer widths.

Given the potential distribution $\phi_{iDG}(z)$, we can, using Pao-Sah's [5] formula, obtain the analytical expression for the drift-diffusion current in the DG JL MOSFET channel as a function of the relevant parameters. In the symmetric case, the total current in the DG JL MOSFET channels can be written as

$$I_{DG} = \frac{eD_n W_x n_i}{L} \left[\int_0^{z_{if}} e^{\phi_{if}(z)/\phi_T} dz + \int_{z_{if}}^{t_{Si}/2} e^{\phi_{pd}(z)/\phi_T} dz + \int_{t_{Si}/2}^{t_{Si}-z_{ib}} e^{-\phi_{nd}(z)/\phi_T} dz + \int_{t_{Si}-z_{ib}}^{t_{Si}} e^{-\phi_{ib}(z)/\phi_T} dz \right] \times \int_0^{U_D} e^{(-\eta(y)-\phi_{Fs})/\phi_T} d\eta. \tag{55}$$

All integrals in this equation can be performed analytically, and for $z_{if} + z_{ib} < t_{Si} < w_p + w_n + z_{if} + z_{ib}$, and $z_{if} = z_{ib}$, the current becomes

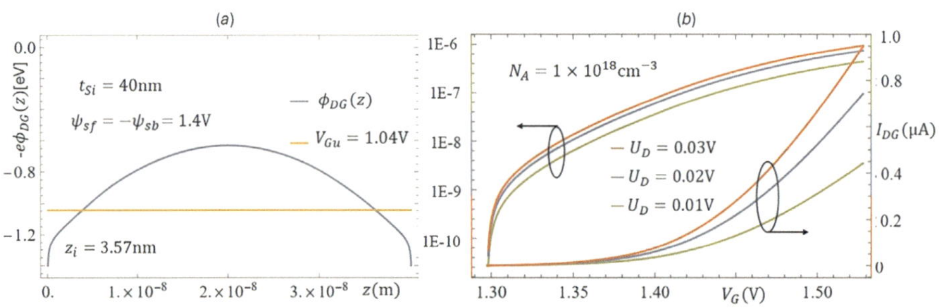

Figure 13.
Electric potential of drift-diffusion current in a symmetric double-gate JL MOSFET. In (a) the electric potential for a DG MOS where the silicon thickness t_{Si} is smaller than the expected depletion plus inversion layers width $w_p + w_n + z_{if} + z_{ib}$. Some parameters are assumed equal to those in Ref. [23], i.e., $L = 11\mu m$, $W_x = 10\mu m$. However, nowhere in Ref. [23] are the impurity concentration and mobilities given. In this reference, the silicon layer width t_{Si} is assumed equal to 8 nm, slightly larger than the sum of the inversion layer widths for $N_A = 1 \times 10^{18}\,cm^{-3}$. Here, we assume that $t_{Si} \simeq 40nm$.

$$
\begin{aligned}
I_{DG} &= \frac{eD_n w_x n_i}{L}\left(1 - e^{-U_D/\phi_T}\right)\left[\frac{4\phi_T}{\xi} e^{(\phi_s - V_{Gu})/2\phi_T} \sinh\frac{\xi z_i}{2\phi_T}\right. \\
&\left. + \frac{\sqrt{\pi\varepsilon_s\phi_T}}{\sqrt{eN_d}} e^{-\phi_F/2\phi_T}\left(\mathrm{Erf}\left[\frac{\sqrt{eN_d}w_p}{2\sqrt{\varepsilon_s\phi_T}}\right] + \mathrm{Erf}\left[\frac{\sqrt{eN_d}\left(t_{Si} - w_p - 2z_{if}\right)}{2\sqrt{\varepsilon_s\phi_T}}\right]\right)\right] \\
&= I_{DGi} + I_{DGd}.
\end{aligned} \tag{56}
$$

It can easily be seen that the current in the DG JL MOSFET is practically twice the charge and current in the single gate MOSFET channel [25]. In fact, it is exactly twice when $t_{Si} = 2w_p + 2z_{if}$.

In **Figure 13**(b) we plot the drift-diffusion current of Eq. (56) as a function of the gate potential, for three different values of the source-drain voltage U_D. For this graph we assume that the impurity concentration is $N_A = 1 \times 10^{18}\ \mathrm{cm}^{-3}$, the channel length is $L = 11\mu\mathrm{m}$, the gate width is $W_x = 10\mu\mathrm{m}$, and the mobility is assumed equal to $40\mathrm{cm}^2/\mathrm{Vs}$. Some of these parameters are the same as those used in Figure 7 of Ref. [23], where a simplified charge distribution is assumed; however, we cannot compare with that reference because some parameters, such as the impurities concentration and mobilities, are not given.

It is clear that the analytical expression for the current in Eq. (51) can also be used for asymmetric DG JL MOSFETs. The general formulas for the inversion layer width, surface electric field, and effective concentration of ionized impurity atoms, introduced for the single gate MOS in Section 5, are also valid with the appropriate changes for the double gate MOSFET. The ability to evaluate these quantities, particularly the widths of the inversion layers, is, for a given impurity concentration, an important input in the design of MOSFETs and in the definition of some device parameters, such as the silicon layer width t_{Si}.

10. Conclusions

We presented here an outline of the new theoretical approach for the analytical calculation of potential distributions and drift-diffusion currents in MOS and JL MOSFETs. We presented a summary of the novel method that makes it possible to solve analytically the nonlinear Poisson equation in the inversion layer of MOSs and MOSFETs. We apply this method to the more precise Kingston-Neustadter charge density model, unsolved until now, where the presence of electrons and holes in the inversion layer is assumed, and also to the most representative of the simplified compact analytical models, the Hauser-Littlejohn model, which assumes that only electrons or only holes are present. The predictions of both models for the potential and charge distributions, as well as for the drift-diffusion current, were compared and we have shown that they generally differ. In the HL model, the widths of the inversion layer are much smaller, and the charge distributions are almost 2D, with a higher surface density and practically insensitive to impurity concentration. The drift-diffusion currents predicted by the HL model are also different and smaller than the currents predicted by the KN model. Independent of these comparisons with the standard approaches, we have presented here several analytical formulas for potential and charge distributions and new analytical formulas for the drift-diffusion current in single and double-gate MOSFETs.

Acknowledgements

I acknowledge valuable comments by Herbert P. Simanjuntak, Jürgen Reiter, and A. Robledo-Martinez.

Conflict of interest

The author declares no conflict of interest.

Author details

Pedro Pereyra[1,2]

1 Area de Física Teórica y Materia Condensada, Universidad Autónoma Metropolitana, Azcapotzalco, Mexico City, Mexico

2 Member of the Sistema Nacional de Investigadores, CONAHCyT, México

*Address all correspondence to: ppereyra@azc.uam.mx

References

[1] Shockley W. The theory of p-n junctions in semiconductors and p-n junction transistors. Bell System Technical Journal. 1949;**28**:435

[2] Kingston RH, Neustadter SF. Calculation of the space charge, electric field, and free carrier concentration at the surface of a semiconductor. Journal of Applied Physics. 1955;**26**:718-720

[3] Kahng D, Atalla MM. Silicon-silicon dioxide field induced devices. Solid State Device Research Conference. Pittsburgh. US3204160A. 1960

[4] Sah CT. Characteristics of the metal-oxide-semiconductor transistors. IEEE Transactions on Electron Devices. 1964; **ED-11**:324-345

[5] Pao HC, Sah CT. Effects of diffusion current on characteristics of metal-oxide (insulator)-semiconductor transistors. Solid State Electonics. 1966;**9**:927-937

[6] Grove AS. Physics and Technology of Semiconductor Devices. New York: John Wiley; 1967. ISBN: 978-0-471-32998-5

[7] Bassett RK. MOS technology, 1963-1974: A dozen crucial years. The Electrochemical Society Interface. Fall 2007. 1970:46-50

[8] Hauser JR, Littlejohn MA. Approximations for accumulation and inversion space-charge layers in semiconductors. Solid State Electonics. 1968;**11**:667-674

[9] Demoulin E, van De Wiele F. Inversion layer at the interface of Schottky diodes. Solid State Electonics. 1974;**17**:825-833

[10] Temple VAK, ShewChun J. The exact low-frequency capacitance-voltage characteristics of metal-oxide-semiconductor (MOS) and semiconductor-insulator-semiconductor (SIS) structures. IEEE Transactions on Electron Devices. 1974;**18**:235-242

[11] Brews JR. A charge-sheet model of the MOSFET. Solid State Electonics. 1978;**21**:345-355

[12] Flandre D, van de Wiele F. A new analytical model for the two-terminal MOS-capacitor on SO1 substrate. IEEE Electron Device Letters. 1988;**9**:296-299. DOI: 10.1109/55.722

[13] Schubert M, Höfflinger B, Zingg RP. An analytical model for strongly inverted and accumulated silicon films. Solid State Electonics. 1990; **33**:1553-1567

[14] Taur Y. An analytical solution to a double-gate MOSFET with undoped body. IEEE Electron Device Letters. 2000;**21**:245-247

[15] Taur Y. Analytic solutions of charge and capacitance in symmetric and asymmetric double-gate MOSFETs. IEEE Transactions on Electron Devices. 2001; **48**:2861-2869

[16] Ortiz-Conde A, García FJ, Muci SJ. Rigorous analytical solution for the drain current undoped symmetric dual-gate MOSFETs. Solid State Electonics. 2005; **49**:640-647

[17] Taur Y, Liang X, Wang W, Lu H. A continuous, analytic drain-current model for DG MOSFETs. IEEE Transactions on Electron Devices. 2004; **25**:07

[18] Roy AS, Sallese JM, Enz CC. A closed-form charge-based expression for drain current in symmetric and

asymmetric double gate MOSFET. Solid State Electronics. 2006;**50**:687

[19] Moreno-Pérez E et al. An inversion-charge analytical model for square gate-all-around MOSFETs. IEEE Transactions on Electron Devices. 2011;**58**:2854

[20] Hu M-C, Jang S-L. An analytical fully-depleted SOI MOSFET model considering the effects of self-heating and source/drain resistance. IEEE Transactions on Electron Devices. 1998; **45**:797

[21] Reyboz M, Rozeau O, Poiroux T, Martin P, Jomaah J. An explicit analytical charge-based model of undoped independent double gate MOSFET. Solid State Electronics. 2006;**50**:1276

[22] Oproglidis TA et al. Analytical drain current compact model in the depletion operation region of Short-Channel triple-gate Junctionless transistors. IEEE Transactions on Electron Devices. 2017; **64**:66

[23] Khandelwal S et al. BSIM-IMG: A compact model for ultrathin-body SOI MOSFETs with Back-gate control. IEEE Transactions on Electron Devices. 2019; **59**:2012

[24] Tsividis Y. Operation and modelling of MOS transistor. In: The Oxford Series in Electrical and Computer Engineering. 2nd ed. Oxford, UK: Oxford University Press; 2003. See Appendix F, and references therein

[25] Pereyra P. Submitted to Semiconductor Science and Technology

[26] Pereyra P. DOI: 10.22541/au.170864077.78820151/v1

[27] Sze SM. Physics of Semiconductor Devices. Singapore: Wiley Interscience; 1981

[28] Shockley W. Electrons and Holes in Semiconductors. New York: D. Van Nostrand; 1950

[29] Shalimova KV. Física de Semiconductores. Moscou: MIR; 1975

[30] Hook JR, Hall HE. Solid State Physics. Chichester West Sussex, England: J. Wiley & Sons; 1991. ISBN: 0-471-92805-4

[31] Seeger K. Semiconductor Physics, an Introduction. 6th ed. Heidelberg, Germany: Springer; 1997. ISBN:3-540-61507-5

[32] Reiter J. Private communication, work in progress

Chapter 5

Various Approaches for Nano-Scale Devices Design

Indra Vijay Singh

Abstract

5G is poised to significantly contribute to meeting the soaring demands for wireless networks, surpassing current standards in quality, efficiency, and speed. The existing bandwidth used by 4G and its predecessors is overwhelmed by the multitude of devices. To counter this, 5G adopts higher frequencies, such as the mm-Wave spectrum, introducing a new dimension for applications across imaging, sensing, radar, and communication. This study delves into various device design strategies like thin body, graded channel, halo doping, and multiple-gate. These approaches are examined in relation to threshold voltage (V_{TH}) and RF performance metrics like A_v, f_T, and f_{MAX}, particularly in the low moderate inversion region. The goal is to cater to low power consumption and high-frequency applications. The findings strongly indicate that sub-100 nm FD-SOI MOSFETs will emerge as formidable contenders for analog/RF applications in the lucrative 5G wireless communications market of today.

Keywords: SOI, MOSFET, underlap, 5G, high gain

1. Introduction

In order to realize higher-speed and higher-packing density in MOS integrated circuits, the dimensions of MOSFETs have continued to shrink according to the scaling law proposed by Dennard et al. [1]. Yet, the power consumption of modern VLSI's has become rather significant issue and reducing this power is strongly desired. Choosing a lower power supply voltage is an effective method. However, it leads to the degradation of MOSFET current driving capability [1]. Consequently, scaling of MOS dimensions is very important in order to improve the drivability, and to achieve higher-performance and higher-functional VLSI's [2].

With aggressive technology scaling to enhance performance, circumventing the detrimental short-channel effects (SCEs) to improve the device reliability has been the focus in MOSFET scaling. When the channel length shrinks, the controllability of the gate over the channel depletion region reduces due to the increased charge sharing from source/drain. SCEs lead to several reliability issues including the dependence of V_{TH} upon channel length. This leads to the scatter of device characteristics because of the scatter of gate length produced during the fabrication process [2].

The predominantly reliability problems associated with SCEs such as pinch off and a shift in V_{TH} with decreasing channel length as well as drain induced barrier lowering

(DIBL) and hot-carrier effect with increasing drain voltage, as a result degrades the linearity of the device which is most important issue for the wireless system [3]. Moreover, SCEs degrades the controllability of the gate voltage to drain current, which leads to the degradation of the sub-threshold slope and the increase in off-current. Thinning gate oxide and using shallow source/drain junctions are known to be effective ways of preventing SCEs. But with thinner gate oxide gate current tunneling phenomena will occur, which is not required [3].

The above description can be applied to conventional MOSFETs fabricated in a bulk silicon wafer. What about SOI MOSFETs? They are attractive devices for low-power high-speed VLSI applications because of their small parasitic capacitance and improved SCEs [4]. Young [5] analyzed the SCEs using a device simulator, and concluded that SCEs is well suppressed in SOI MOSFET's with compared to bulk MOSFET's [3].

In general, it is believed that SOI-MOSFETs have a higher immunity to SCEs compared with bulk MOSFETs. This may be due to the difference in source/drain junction depths between the two kinds of devices. For instance, the thickness of the silicon film, T_{si} which corresponds to the source/drain junction depth of 50–100 nm, is common in 0.25–0.35 μm SOI MOSFET's. It is extremely shallow compared with the junction depth of 100–200 nm in 0.25–0.35 μm gate bulk MOSFET's. However, to take advantage of the ameliorated SCEs in sub 100 nm FD-SOI, T_{si} must be considerably smaller and should be one third to one fourth of gate length ($T_{si} \cong 30$ to 25 nm) as described in [3, 6]. Moreover, a strong coupling through the buried oxide in thin-film devices exists and consequently, very thin buried oxides ($T_{BOX} \cong 100$ nm) are needed which trade-offs with junction capacitance considerations. With the gate length scaling approaching sub-100-nm regime, a optimized body-doping concentration is required for improving performance and integration density.

Gate oxide thickness, and source/drain (S/D) doping profiles to control SCEs become increasingly difficult to meet when abrupt doping profile used in conventional device structures based on bulk silicon substrates are employed. Moreover SOI brings in new reliability issues, which are not known in the traditional bulk-Si devices, related to the presence of the BOX like self-heating and hot-electron degradation of the BOX [7]. In a high electrical field of a short-channel transistor, carriers may gain enough energy and get trapped in the BOX along with the gate oxide and does not reach in to the substrate therefore, no body effects phenomenon will occur.

The presentation of this chapter is organized as follows: Section 2 presents an overview of various design approaches for FD-SOI MOSFETs such as thin body, graded channel, hallo doped and multiple-gate design. The gate-underlap also known as source/drain-extension (SDE) device design engineering has been elaborated in Section 3. Furthermore, investigations of SDE engineering in-terms of natural length, threshold voltage and optimum silicon thickness have been presented in Section 4. Various steps involved in the design of gate-underlap MOSFETs have been given in Section 5. In addition, the device design has been verified through simulation for dc-IV characteristics and effects of process parameters on threshold voltage V_{TH}, intrinsic gain A_v, f_T, f_{MAX} and finally conclusions are taken up.

2. Various design approaches for FD-SOI

A FD-SOI MOSFETs is a good candidate for RF IC application because of its low cost, low power and high integration potential [2]. It has the advantages of lower junction capacitance and better sub-threshold swing. The performance parameters of device for high frequency analog applications such as high f_T, f_{MAX}, high g_m, low g_{ds}, low junction capacitance, and small V_{TH} variations are required. The most important among them are g_m and g_{ds}, because they determine the intrinsic gain A_v (g_m/g_{ds}) of an amplifier. For improving the SCEs (small V_{TH} variation) and analog/RF performance of FD-SOI MOSFETs in nano-scale regime for low power GHz range application, many device design approaches have been adopted, which are summarized below:

2.1 Thin body

In order to realize high packing density, high cut-off frequency and low noise MOSFETs for future ULSIs design both the channel length and width must be reduced [5]. A decrease in channel length leads to SCEs, whereas reduction in channel width causes a reduction in drive current and reliability degradation due to high electric field. Therefore, for maximizing the gate electrostatic control over the entire depth of the body thickness, reducing of SCEs as a result eliminates the surface leakage paths in FD-SOI MOSFETs, requires the use of thin silicon films. A device structure that implements this concept is the thin-body MOSFET, which is shown in **Figure 1(a)** [5].

In thin-body MOSFET, the source-to-drain current is restricted to flow in a region close to the gate for superior gate control. Since it does not rely on a heavily-doped channel for the suppression of SCEs, it avoids the problems of lower mobility degradation due to impurity scattering and V_{TH} fluctuation due to the random variation of the number of dopant atoms in the channel region of nano-scale transistors [5]. But thin body SOI-MOSFETs has thin source/drain (S/D) region would contribute a high series resistance that degrades the drive current (I_{on}) as shown in **Figure 2(a)** and **(b)**, respectively. Due to degradation in I_{on}, transconductance g_m of the device will degrade; as a result intrinsic gain would decrease. Furthermore, due to contribution of high series resistance, analog/RF performance metric f_{MAX} would be decreased.

Therefore, for avoiding the problem of thin source/drain (S/D) region, a raised S/D is introduced (see **Figure 1(b)**) to avoid the series resistance and I_{on} problems [5, 8]. Nevertheless, parasitic capacitances between the raised source/drain (S/D) and the

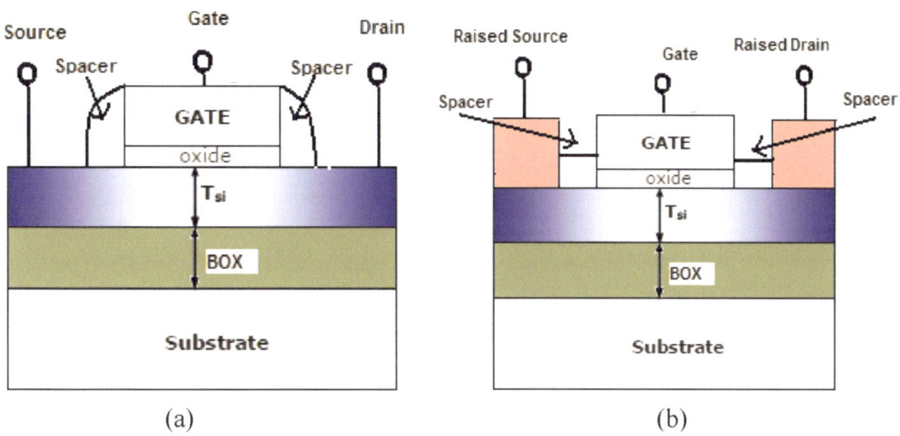

Figure 1.
Comparison of device structure for (a) a conventional thin body FD-SOI MOSFETs and (b) a raised source/drain FD-SOI thin-body MOSFETs.

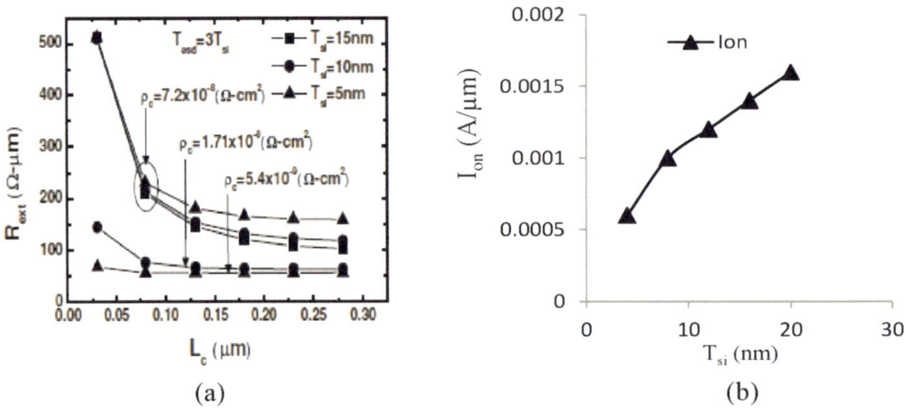

Figure 2.
Plot of (a) external resistance R_{ext} vs. contact length L_c with T_{si} = 15, 10 and 5 nm, respectively [8] (b) T_{si} dependence of I_{on} for thin body device [5].

gate are inherent in this type of devices structure. This is expected to adversely impact the device speed and power consumption [5]. Elevated S/D structure reduce the series resistance, but the much higher S/D resistance and parasitic capacitance than the expected have still been observed in the experimentally, demonstrated [8]. Therefore, thin body fully depleted SOI-MOSFETs may not be suitable technology for high frequency application.

2.2 Graded channel (GC)

In the graded channel FD-SOI MOSFETs as shown in **Figure 3**, the asymmetric channel profile is obtained by a simple mask arrangement system [9]. In this type of device, high concentration at the source end (L_{HD}) improves the V_{TH} roll-off and (DIBL), while the low doping near the drain (L_{LD}) ensures high mobility, reduced peak electric field and impact ionization has been kept as discussed in [9, 10]. The GC device can be viewed as two sub-devices connected in series: an enhancement mode device at the source side and a depletion mode device at the drain side, each having a different V_{TH} and a different "channel" length. The effective channel length (L_{eff}), when the device is "on", is mainly determined by the P-doped region, which is much shorter than the physical gate length. As the result, for the same physical drawn gate length, the GC device can provide higher drive current, higher peak transconductance, higher output resistance than a conventional FD-SOI device with uniformly doped channel [10].

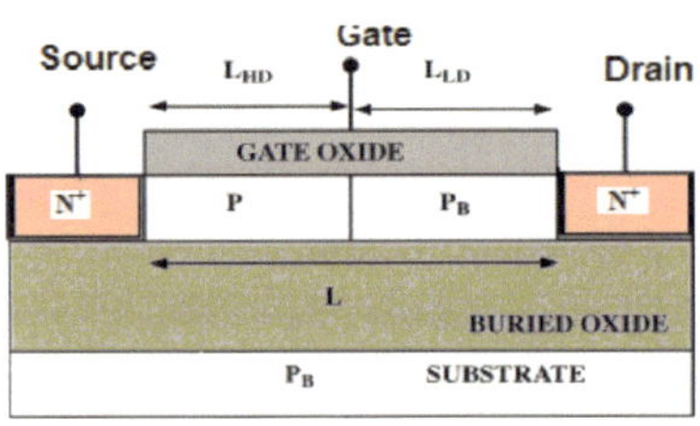

Figure 3.
Cross section of a graded channel (GC) FD-SOI nMOSFETs. P-implemented boron concentration in the high doped region (L_{HD}): P_B-starting boron concentration in the low doped region and substrate (L_{LD}).

Name	V_{GS} (V)	V_{DS} (V)	f_T (GHz)	g_m (mS)	g_{ds} (mS)	C_{gs} (fF)	C_{gd} (fF)	g_m/g_{ds}	C_{gs}/C_{gd}
Conventional FD-SOI MOSFETs, L_G = 0.25 μm	1	1.5	28	11.5	1.8	32	14.5	6.38	1.5
GC FD-SOI MOSFETs L_G = 0.5 μm, L_{HD} = 0.25 μm	1	1.5	19	9.8	0.3	62	20	32.7	3.1

Table 1.
Extracted values of the equivalent circuit of a conventional MOSFET and GC-MOSFET [11].

The technological aspects and detail dc simulation of GC FD-SOI MOSFETs were presented in [10]. The RF performances and potentialities of GC MOSFET have been demonstrated using various figures of merit and compared with conventional FD-SOI MOSFETs in [11]. **Table 1**, provides extracted values of g_m, g_{ds} and C_{gs} of a GC and conventional FD-SOI MOFETs as discussed in [11].

As discussed in the previous section that g_m and f_T are the most important factor for the RF performances of MOSFET, from the **Table 1** it can be seen that both (g_m and f_T) are lower in GC-MOSFETs compared to conventional MOSFETs. Therefore, GC is not very suitable device technology for GHz applications.

2.3 Halo-doped

With continuous device scaling down to 100 nm channel length and less, the halo (or pocket) implantations have been introduced to better control of the SCEs shown in **Figure 4** [12, 13]. In digital applications halo implantations have the purpose of reducing the off state leakage current while maximizing transistor linear and saturated drive currents. While for analog applications it has been shown that halo implantation is needed for base-band applications using longer channel, it has detrimental effect for high speed applications using minimum channel transistors in strong inversion region [13]. Excessive halo implantation in PD-SOI transistors increases the kink effect. Halo implantation is also known to degrade the distortion characteristics when the SOI devices are used as resistors [14].

The parameters for high performance analog applications are high f_T, f_{MAX}, high g_m, low g_{ds}, low junction capacitance, and small V_{TH} variations as discussed previously. The most important among these are g_m and g_{ds} because they determine the intrinsic gain A_v (g_m/g_{ds}) of an amplifier as discussed above. A cross section, of asymmetric single halo (SH) MOSFET structures is shown in **Figure 4** [12].

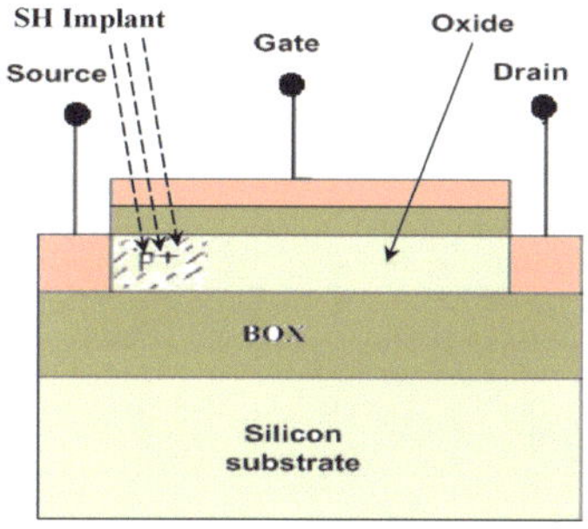

Figure 4.
Cross-section of a single halo (SH) SOI nMOSFET [12].

The details of device design and performance have been discussed in [12]. In this engineering heavy p-type doping near the source, and much light p-type doping near the drain were done [14]. The heavy p-type doping near the source absorbs a higher potential drop, as a result that electric fields near the drain are lower. The lower value of peak electric field near the drain also improves the hot-carrier immunity of SH MOSFETs and reduced SCEs. These types of devices were fabricated by SIMOX process as discussed in Chapter 2. Standard CMOS technology has been used for the fabrication. For the conventional MOSFETs the channel implant was carried out before the gate oxidation, whereas the implant for the SH SOI devices was carried out after the gate formation, at different tilt angles of 7°, 10° and 15° and gate dimensions were defined by electron-beam lithography as discussed in [15].

This type of design adjusts the V_{TH} and improves the device's SCEs as well as reduces the hot carrier effects (HCEs). These devices also achieve higher drive currents by exploiting the velocity overshoot phenomenon [15], which is an advantageous for mixed mode analog/digital circuits. It has been shown that these devices show a marginal improvement in g_m whereas very small value of g_{ds} decreases as compared to the conventional SOI devices. Therefore, in SH devices marginally improved A_v is obtained. The other advantages of SH devices over conventional SOI like absence of kink, lower inherent parasitic bipolar junction transistor (PBJT), gain have also been reported [14]. As discussed in [12] it has been found that single halo doped device technology exhibits pronounced non-quasi-statics (NQS) effects for GHz range frequency applications. Therefore, this type of devices is not fit for high frequency applications.

2.4 Multiple-gate MOSFETs

To prevent the encroachment of electric field lines from the drain to the channel region, special gate structures can be used name as multiple gate as shown in **Figure 5**. Such "multiple"-gate devices include double-gate transistors, triple-gate devices such as the quantum wire [17], the FinFET [18] and gate all around (GAA) [17], the DELTA transistor, and vertical pillar MOSFETs [16, 18]. The multiple gate devices achieve higher values of g_m than single gate (SG) devices at lower gate biases due to the better volume inversion effects [18].

Figure 6 shows the DIBL and threshold voltage roll-off as a function of gate voltage for double, triple, gate all around (GAA) and Π-gate devices. The best performance is obtained from the gate all around, but Π-gate is close second. The results show the efficient shielding of the channel by the gate electrode from the electric field lines originating from the drain region.

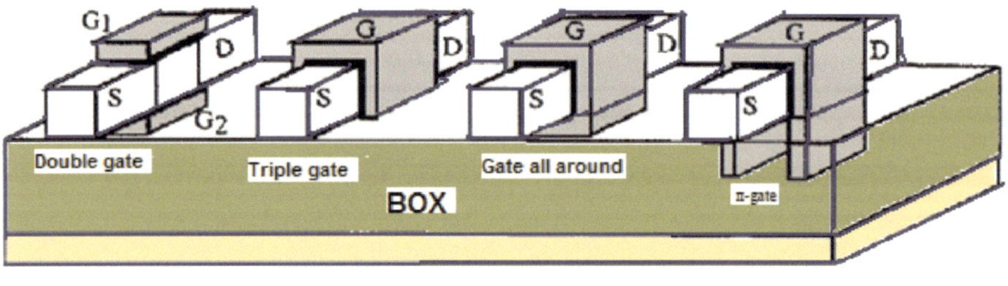

Figure 5.
Double gate, triple gate, gate all around (GAA) and π-gate SOI MOSFETs [16].

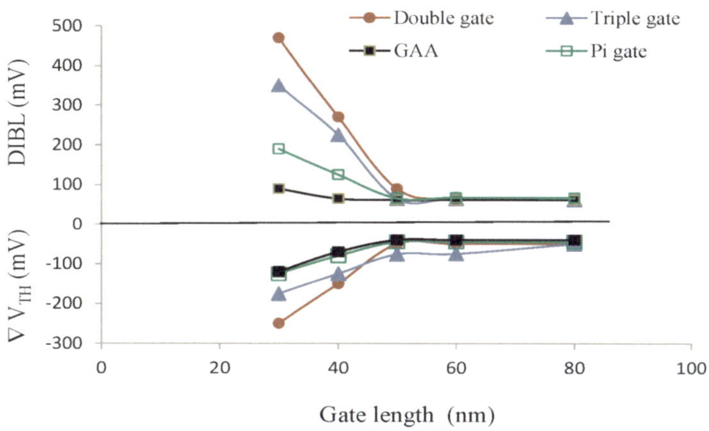

Figure 6.
V_{TH} roll-off and DIBL in double, triple, gate all around (GAA) and π-gate SOI-MOSFETs W = T_{si} = 30 nm [17].

But for high frequency application, required high g_m, f_T, lower g_{ds} and low value of device inherent capacitance. Therefore, we have examined from literature the value of g_m and total gate capacitance C_{GG}, for these devices where f_T is defined as $f_T = \frac{g_m}{2\pi C_{GG}}$ [18]. The ability of device architecture to achieve higher values of f_T depends on achieving a higher g_m along with a reduced value of C_{GG}. From.

Figure 7 it can be seen that multiple gate devices (Triple gate FinFET) exhibits superior value of g_m, but have higher value of C_{GG} compared to SG and DG devices. Whereas f_T values for SG devices at same bias are nearly 40% and 5% higher than those of Triple gate FinFET and DG MOSFETs, respectively [17]. Therefore, it can be concluded from literature study that multiple gate devices, fabricated in non–planar technology exhibit higher parasitic which degrade analog/RF performance metrics [18]. Therefore SG and DG-SOI MOSFETs are possible candidates for device scaling at the end of ITRS roadmap and regarded as next generation of VLSI circuits [19, 20].

Above various device approaches have been discussed, which are reported in literature. From the above discussions, it is obvious that these devices exhibit a lot of problem for high frequency applications such as higher parasitic capacitance, lower g_m and f_T as well as pronounced NQS effects in GHz frequency range. These limitations can be safely overcome by gate-underlap (SDE) design which has lower inherent

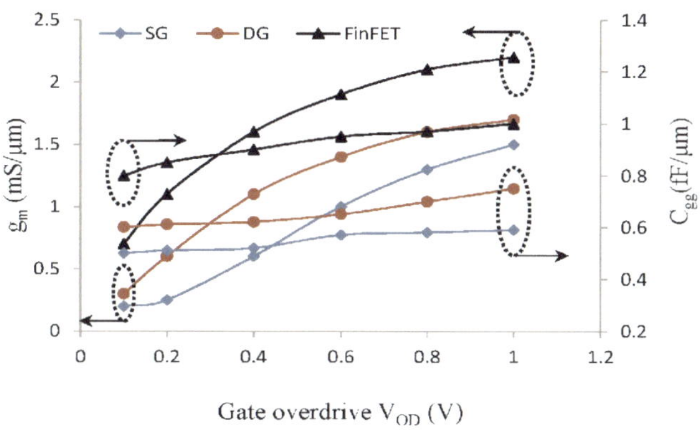

Figure 7.
Dependence of the g_m and C_{GG} for various devices on V_{OD} [17].

parasitic capacitance, improved SCEs, higher g_m (VOD < 100 mV) and f_T. Therefore, SDE design of MOSFETs has become a popular candidate for high frequency low power applications. Details about the SDE device design is elaborated in the next section.

3. Source/drain-extension (SDE) device design

Conventionally, MOSFETs are designed such that source/drain profiles extend beneath the gate known as abrupt doping profile, resulting in gate-source/drain overlap between gate and S/D extension regions. This leads to an overlap capacitance (C_{ovl}) and a reduction in effective gate length (L_{eff}). While C_{ovl} degrades f_T and shorter L_{eff} results in loss of gate controllability leading to SCEs as a result reduced "on" current I_{on} and g_m [17]. These parameters limit the scaling of analog MOS devices for high frequency applications and should be addressed to enable downscaling. Therefore, in addition to change the transistor architecture, channel engineering has become an attractive option to enhance the device performance. One such concept of channel engineering is the use of "gate-underlap" also known as source/drain extension region (SDE) engineering [21–25]. In gate-underlap architecture, source/drain extension region profile are designed such that extension region doping does not approach the gate edge i.e. the channel and extension regions adjacent to the gate will be without any dopant [3].

Recently, the use of gate-underlap doping profile in nano-scale MOSFETs has attracted attention due to its ability to suppress SCEs, reducing fringing field and improving analog/RF metrics for low voltage high frequency applications [3, 7]. This technique improves the analog performance matrices like A_v, f_T, g_m and g_m/I_{DS}. The g_m/I_{DS} ratio (where I_{DS} represents drain-to- source current) indicate the efficiency of the devices to convert dc power in to ac frequency gain performance. Another important issue in nano-scale regime to control the dopant profile at the source end of the channel which can be control using gate-underlap technique. The detail SDE engineering was elaborated in next section.

3.1 What is source/drain extension (SDE) engineering?

In SDE engineering highly doped source and drain region is away from the gate edge by a distance 's' known as spacer [4]. The Gaussian doping profile along with lateral source/drain doping gradient d nm/decade is used, instead of abrupt doping profile used in conventional design [4]. Undoped thin silicon film is used for enhanced channel mobility, avoids random dopant fluctuation and reduces SCEs as well as achieved low I_{off}.

This device design technology is more beneficial in low moderate inversion region (suitable for low power applications) [26]. In this kind of design with highly doped SDE regions extra series resistance can be minimized to achieve higher "on" current I_{on}. This source/drain engineered device is very advantageous from analog perspective, due to the bias dependent effective gate length L_{eff} and it is related by s and lateral doping gradient d.

In this design technique for larger value of s and lower value of d, L_{eff} will be larger than physical gate length L_G ($L_{eff} > L_G$) [26]. However, larger value of s provides higher series resistance and reduces I_{on} while lower value of d corresponds to an abrupt phenomenon which is hard to realize in actual fabrication scenario. On other hand smaller value of s and higher value of d result in L_{eff} smaller than L_G ($L_{eff} < L_G$), which results in higher drive current, higher peak transconductance, higher output resistance than a conventional (abrupt) design. But smaller value of s contributes higher inherent capacitance whereas higher value of d would cause significant fluctuation in the dopant underneath the channel resulting in undesirable effects. Therefore, for optimal performance of nano-scale SOI MOSFETs careful engineering of SDE regions, (which are determined by d and s) is required. This result in improved SCEs and reduced parasitic effects in nano-scale SOI-MOFETs for low power GHz applications [27–29].

The FD-SOI MOSFETs has received great attention in recent years owing to the inherent suppression of short-channel effects (SCEs), excellent value of sub-threshold slope (S), improved value of I_{on} and g_m with low value of parasitic as discussed in Chapter 2. However, a strong demand of lower V_{TH} to improve the I_{on} for high-speed operation, leads to an increase in off-current (I_{off}) for nano-scale FD configurations. This further leads to a negative consequence of increase in stand-by power consumption in integrated circuits designed with nano-scale SOI MOFETs. Therefore, remedy for this I_{off} can be reduced by an efficient control on SCEs by reducing the silicon film thickness T_{si}. However, as the gate length is reduced below 100 nm, fabrication of ultra-thin defect free T_{si} remains a technological challenge [4]. Reducing the T_{si} introduces an external parasitic resistance in series with the channel that degrades the I_{on}. Therefore, the optimization of s, d and T_{si} for SCEs immunity must be carefully considered for high I_{on}/I_{off} current ratio. By increasing doping gradient for a low value of spacer, (with given T_{si}) at the gate edge not only reduces L_{eff} (increase SCEs) but it also draws more carriers in to the spacer region near the gate edges, thus it increases I_{on}. Therefore, lateral source/drain doping gradient d along with the spacer s and silicon film thickness T_{si} presents a design trade-off between acceptable SCEs and parasitic series resistance for low power and high frequency applications. Therefore, in this kind of scenario the source/drain extension (SDE) regions must be carefully engineered for the overall improved high frequency performance of a nano-scale device. The analytical SDE device theory has been presented in next section.

3.2 Theory of SDE

A detail insight of FD-SOI MOSFET, (with lateral source/drain doping gradient d, and spacer s) shown in **Figure 8** can be obtained by solving the following 2D Poisson equation in the silicon film [4, 5].

$$\frac{\partial^2\psi(x,y)}{\partial x^2} + \frac{\partial^2\psi(x,y)}{\partial y^2} = \frac{q}{\varepsilon_{si}}\left[N_A + N_{SD}e^{\frac{x^2}{\sigma^2}} - N_{SD}\frac{\left(L_g + 2s - x\right)^2}{\sigma^2}\right] \tag{1}$$

Where $\psi\ (x,y)$ is the 2D potential in the silicon film, N_A is the silicon film doping, N_{SD} is the peak doping concentration in the source/drain region, ε_{si} is the dielectric permittivity of the silicon and σ is the parameter governing the lateral source/drain doping gradient (d) and the spacer (s) and it is defined as [4].

$$\sigma = \sqrt{\frac{2sd}{\ln(10)}} \tag{2}$$

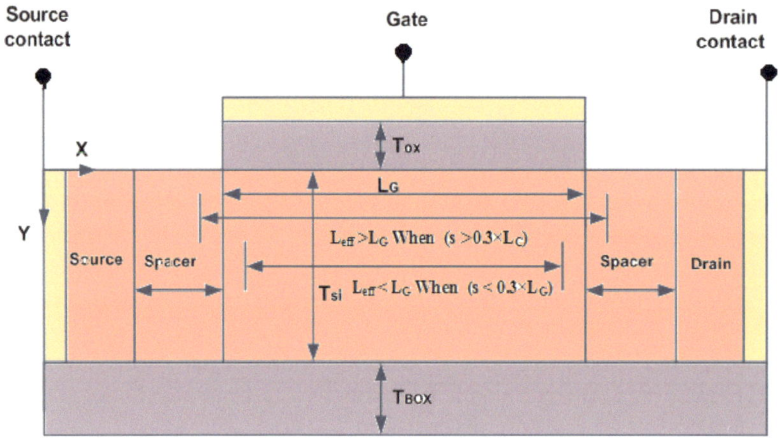

Figure 8.
Cross-section of FD-SOI MOSFET. Two different values of the effective gate length (L_{eff}) are shown to illustrate that L_{eff} can be longer or shorter than the gate length (L_G) depending on the spacer s and source/drain doping gradient d.

σ is in the exponential term of (1), it signifies the roll-off of the source/drain Gaussian profile in the lateral direction. Its value is selected to obtain the desired d at a given spacer, which is given by (3), where N_D is donor type doping at point x in the spacer.

$$d = \left| \frac{d\ \log(N_D(x))}{dx} \right|^{-1} \tag{3}$$

The exact analytical solution of (1) is mathematically very complex and may not be suitable for implementation in a compact model. Therefore, we approximate the impact of s and d by introducing the concept of an effective gate length (L_{eff}) and solve effective analytical expression for L_{eff} as a function of spacer s, doping gradient d, source/drain doping level and gate length. Under these assumptions, (1) is solved.

without the terms ($N_{SD}e^{\frac{x^2}{\sigma^2}}$) and ($N_{SD}\frac{(L_g+2s-x)^2}{\sigma^2}$). The final expression may be approximated as below:

$$\frac{\partial^2\psi\,(x,y)}{\partial x^2} + \frac{\partial^2\psi\,(x,y)}{\partial y^2} = \frac{q}{\varepsilon_{si}}N_A \tag{4}$$

In the present analysis, we use superposition principle to solve the 2D Poisson equation (4) without any approximations (such as parabolic profile along the vertical direction) for potential distribution in the silicon film. The 2D potential $\psi\,(x,y)$ can be split in to two part using superposition, a long channel solution to 1D Poisson equation. U(y) and the other a short channel solution to 2D Laplace equation. V(x,y).

$$\psi\,(x,y) = U(y) + V(x,y) \tag{5}$$

The boundary condition is used for the solution of (4) are given as:

$$\left.\frac{\partial\psi\,(x,y)}{\partial y}\right|_{y=0}=\frac{C_{ox}}{\varepsilon_{si}}\left(\psi\,(x,y)\rfloor_{y=0}-V'_{GS}\right) \tag{6}$$

$$\left.\frac{\partial\psi\,(x,y)}{\partial y}\right|_{y=T_{si}}=-\frac{C_{ox}}{\varepsilon_{si}}\left(\psi\,\left(x,y\right|_{y=T_{si}}-V'_{GS}\right) \tag{7}$$

$$\psi(x,y)\left.\right|_{x=0}=V_{bi} \tag{8}$$

$$\psi\,(x,y)\left.\right|_{x=L_{eff}}=V_{bi}+V_{DS} \tag{9}$$

Where $C_{ox}\left(=\frac{\varepsilon_{ox}}{T_{ox}}\right)$ is the gate oxide capacitance, $C_{si}=\frac{\varepsilon_{si}}{T_{si}}$ is the silicon film capacitance, $V'_{GS}=V_{GS}-V_{fb}$ with V_{GS} and V_{fb} being the gate to source and the flat band voltages, respectively, V_{DS} is the applied drain voltage and V_{bi} is the build in voltage.

It is important to note that the boundary condition has been evaluated at x = L_{eff} instead of L_G. Then the effective gate length L_{eff} of the device can be modeled as:

$$L_{eff}=L_G+2\left(s-\sigma\sqrt{\ln\left(\frac{N_{SD}}{\eta_{SD}}\right)}\right) \tag{10}$$

The effective gate length L_{eff} has three components gate length (L_G), spacer width (s) and the contribution of lateral doping gradient (d) through the terms σ and η_{SD}, Where η_{SD} represents the source/drain doping level at which the effective gate length is determined. The factor of 2 in the second term represents the contribution from source and drain side extension regions to L_{eff}. The term $\sigma\sqrt{\ln\left(\frac{N_{SD}}{\eta_{SD}}\right)}$ represents the distance from the spacer at which the source/drain doping profile reaches a particular source/drain doping level. The term $\left(s-\sigma\sqrt{\ln\left(\frac{N_{SD}}{\eta_{SD}}\right)}\right)$ signifies the degree of contribution of the extension region to the gate length. η_{SD} is a function of the spacer width which account for the fact that for larger spacer regions, the gate does not control the extension regions effectively [3]. The L_{eff} for gate-underlap device structure is evaluated when the source/drain doping $\left(N\,(x)=N_{SD}e^{\frac{-x^2}{\sigma^2}}\right)$ reaches η_{SD} and not N_A.

As shown in **Figure 8**, the effect of the lateral doping gradient (d) is accounted for the L_{eff} through the parameter σ and from (10) it can be seen that for large source/drain lateral doping gradients (d) and smaller spacer widths (s), the second terms $2\left(s-\sigma\sqrt{\ln\left(\frac{N_{SD}}{\eta_{SD}}\right)}\right)$ becomes negatives, thus implying that L_{eff} is smaller than the physical gate length (L_G) whereas for larger spacers, $L_{eff} > L_G$. It must be clarified that in source/drain engineered devices, the origin (0,0) for x and y coordinates (usually at the gate edge for devices with abrupt source/drain regions) required for the boundary condition (8) is taken at a lateral distance $\pm\left(s-\sigma\sqrt{\ln\left(\frac{N_{SD}}{\eta_{SD}}\right)}\right)$ from the gate edge. As the position of the origin depends on s and d, its location shifts further away from the gate edge for devices with large s and low d values indicate of $\left(s-\sigma\sqrt{\ln\left(\frac{N_{SD}}{\eta_{SD}}\right)}\right)$ that the devices have been designed with large s and low doping i.e. origin is towards the left of the gate edge i.e. $\Delta L = (L_{eff} - L_G)$ is positive, resulting in a larger effective gate length, where as negative values of $\left(s - \sigma\sqrt{\ln\left(\frac{N_{SD}}{\eta_{SD}}\right)}\right)$ signify that the origin is under the gate (towards the drain), resulting in a shorter effective gate length $L_{eff} < L_G$ [3].

Therefore, according to mathematical expression for better performance of gate-underlap device in-terms of reduced SCEs, high I_{on} and small parasitic capacitance, the value of s and d should be optimized.

4. Investigation of SDE design

For low power high frequency analog application V_{TH} of the device should be small and there should be trade-off between parasitic capacitance and series resistance, therefore s and d optimization should be carried out carefully. It is possible to predict how small the silicon film thickness should be in FD-SOI devices to avoid SCEs (or, at least, to maintain a decent subthreshold slope). Subthreshold slope degradation and other SCEs are caused by the encroachment of electric field line from the drain to the channel region, thereby competing for the available depletion charge, and reducing the V_{TH}. Several studies argue that T_{si} in SOI devices must be thinner one third to one fourth of the physical gate length to control the SCEs [18]. It is therefore, not recommended to design devices with silicon film thickness less than that value. Some mathematical insight in this investigation is presented in the next section.

4.1 Natural length

The natural length gives a measure of the SCEs inherent to a particular device structure. In 1992, Yan et al. solved the Poisson equation in single and double gate MOSFETs and introduced a parameter called the "natural length", λ. It represents the penetration distance of the electric field lines from the drain in the body of the device or the amount of control the drain region has on the depletion zone in the channel, as both the gate and the drain compete for that control [30, 31]. According to literature study, a device can be considered free of SCEs if the gate length is at least 5–10 times larger than λ. A year later, Suzuki et al. refined the model and proposed an accurate expression for the natural length [32]. In single gate device the natural length can be calculated using (11) as discussed in [18].

$$\lambda = \sqrt{\frac{\varepsilon_{si}}{\varepsilon_{ox}}\left(1 + \frac{\varepsilon_{ox} T_{si}}{4\varepsilon_{si} T_{ox}}\right) T_{si} T_{ox}} \tag{11}$$

Where T_{si} and T_{ox} are the silicon and oxide thicknesses and ε_{si} and ε_{ox} are the permittivity of silicon and silicon dioxide, respectively. In 1997, Auth et al. calculated the natural length for a surrounding-gate, cylindrical device with a thin gate oxide. It can be expressed as:

$$\lambda = \sqrt{\frac{\varepsilon_{si}}{4\varepsilon_{ox}}\left(1 + \frac{\varepsilon_{ox} T_{si}}{4\varepsilon_{si} T_{ox}}\right) T_{si} T_{ox}} \tag{12}$$

Based on the similarity between these relationships, natural length concept can be generalized to all MuGFETs by writing as (13).

$$\lambda = \sqrt{\frac{\varepsilon_{si}}{n\varepsilon_{ox}}\left(1 + \frac{\varepsilon_{ox}T_{si}}{4\varepsilon_{si}T_{ox}}\right)T_{si}T_{ox}} \tag{13}$$

Where n is called the "equivalent number of gates" (ENG). ENG is basically equal to the number of gates (if a square cross section is assumed). Thus, we have ENG = 1 for a SG-FD-SOI MOSFET, ENG = 2 for a DG-SOI-MOSFETs and ENG = 4 for a quadruple gate MOSFET. Using multi-gate devices, it is possible to trade a thin T_{ox} for thin T_{si} since λ is proportional to the product $T_{si} \times T_{ox}$. A small λ is desired to minimize SCEs on the subthreshold slope [18]. The natural length concept can be used to estimate the maximum silicon film thickness, gate length and device width that can be used in order to avoid SCEs. The natural length decreases when the number of gates is increased. This improvement is due to the increased influences of the gate over the potential in the channel region. It is important to note that the above relationships are independent of the doping concentration. Thus, the "effective number of gates" concept can be applied to doped channels as well as to devices using mid-gap gate material and undoped channels [16].

4.2 Threshold voltage model

The threshold voltage of SOI devices depends on the T_{si}. Considering only film thickness above 10 nm to avoid quantum mechanical confinement effects, and assuming that the surface potential is $2\varphi_F$ at inversion, from the linear relationship between threshold voltage and T_{si}, one can empirically derive the following general threshold voltage law for multi-gate devices as discussed in [18]:

$$V_{THm} = V_{FB} + 2\varphi_F + \frac{qN_A}{C_{ox}}\frac{T_{si}}{n} \tag{14}$$

Where V_{THm} is threshold voltage of multiple gate devices. The equivalent gate number n depends on the gate extension depth for pi-gate devices and the lateral extension depth of the side gate in omega-gate devices.

The threshold voltage of SG-FD-SOI MOSFETs is expressed as:

$$V_{TH} = V_{FB} + 2\varphi_F + \frac{qN_AT_{si}}{C_{OX1}} \tag{15}$$

When N_A is high, V_{TH} is sensitive to T_{si}, which is drawback for thin FD-SOI devices. When channel is undoped, the depletion charges are nearly zero, and then equation of threshold voltage simplifies to:

$$V_{TH} = V_{FB} + 2\varphi_F \tag{16}$$

It is oblivious from (16) that V_{TH} of the transistor is set by the gate work function w and flat band voltage V_{FB}.

4.3 Optimum silicon film thickness

Although the use of very thin silicon films is required for the suppression of SCEs while thin film thickness raises issues of mobility degradation, high source and drain series resistance, resulting in decreased I_{on}, g_m and f_{MAX} [5]. The natural length concepts can be used to estimate the maximum allowed T_{si} that will avoid SCEs and extra parasitic series resistance for low power GHz range applications. In this work it is defined a T_{si} scaling parameter $\propto_n$ that allows one to estimate the short-channel sensitivity of devices for different gate structures.

$$\alpha_n = \frac{L_G}{2\lambda} \tag{17}$$

Where L_G = gate length and λ = natural length, from (13) the value of λ put in (17) then α_n may be represented as:

$$\alpha_n = \frac{L_G}{2\sqrt{\frac{\varepsilon_{si}}{n\varepsilon_{ox}}\left(1 + \frac{\varepsilon_{ox} T_{si}}{4\varepsilon_{si} T_{ox}}\right) T_{si} T_{ox}}} \tag{18}$$

Where n is called the "equivalent number of gates". In order to insure a sub-threshold slope < 75 mV/dec and DIBL <50 mV, α_n needs to be >2.2 for all devices, which corresponds to imposing that the gate length should be larger than five times the natural length λ. Similarly, α_n was estimated 2.7 for acceptable SCEs like DIBL and hot electron effects in SOI devices. For obtaining the value of $\alpha_n > 2.7$, the value of λ should be nearly five-to-ten times smaller than L_G whereas thickness of T_{si} should be nearly three-to-four times smaller than L_G. For a given value of L_G, oxide thickness and α_n, it is possible to determine a minimum T_{si} allowing for SCEs and source/drain extra series resistance problem free operation of devices and can be calculated using (18) as discussed in [33]. In our design α_n was calculated as $\cong 4.7$.

5. Design verification through simulation

5.1 Device design simulation setup

At GHz frequencies, the device design and optimization has a significant impact on overall performance. As a result careful device design becomes quit important in pushing the capability of nano-scale SOI to higher GHz range (>10 GHz). The device design flow charts are given in **Figure 9**.

The device has been designed by followed the steps enumerated in the flow chart (see **Figure 9**). The basic physical structure of underlap SOI configurations comprising gate oxide T_{ox}, thin silicon T_{si}, buried oxide thickness T_{BOX}, gate length L_G spacer s, source/drain doping gradient d, channel doping N_A and mid gap material gate work function w of 4.62 eV are precisely used. In Sections 5.3 and 5.4, these process parameters are optimized for low value of V_{TH} and higher values of A_v, f_T and f_{Max}.

The 90 nm SOI technology used in this work is commercially available from the foundries of, for instance, Fujitsu, Texas Instruments, TSMC, IBM and UMC. The optimized values of these process parameters are found as: the acceptor doping used in Si layer 1×10^{16} cm^{-3} for adjusting a low value of V_{TH} = 0.26 V, which is required for low power applications. Source and drain regions were doped uniformly with donor density of 5×10^{20} cm^{-3}. According to ITRS [20], oxide thickness T_{ox} should be 0.7 nm for controlling the channel charge but here a rather conservative T_{ox} (3 nm) has been used because of the potential problems that may appear during the growth of a very thin oxide on the sidewalls and at the corners of the silicon island [18].

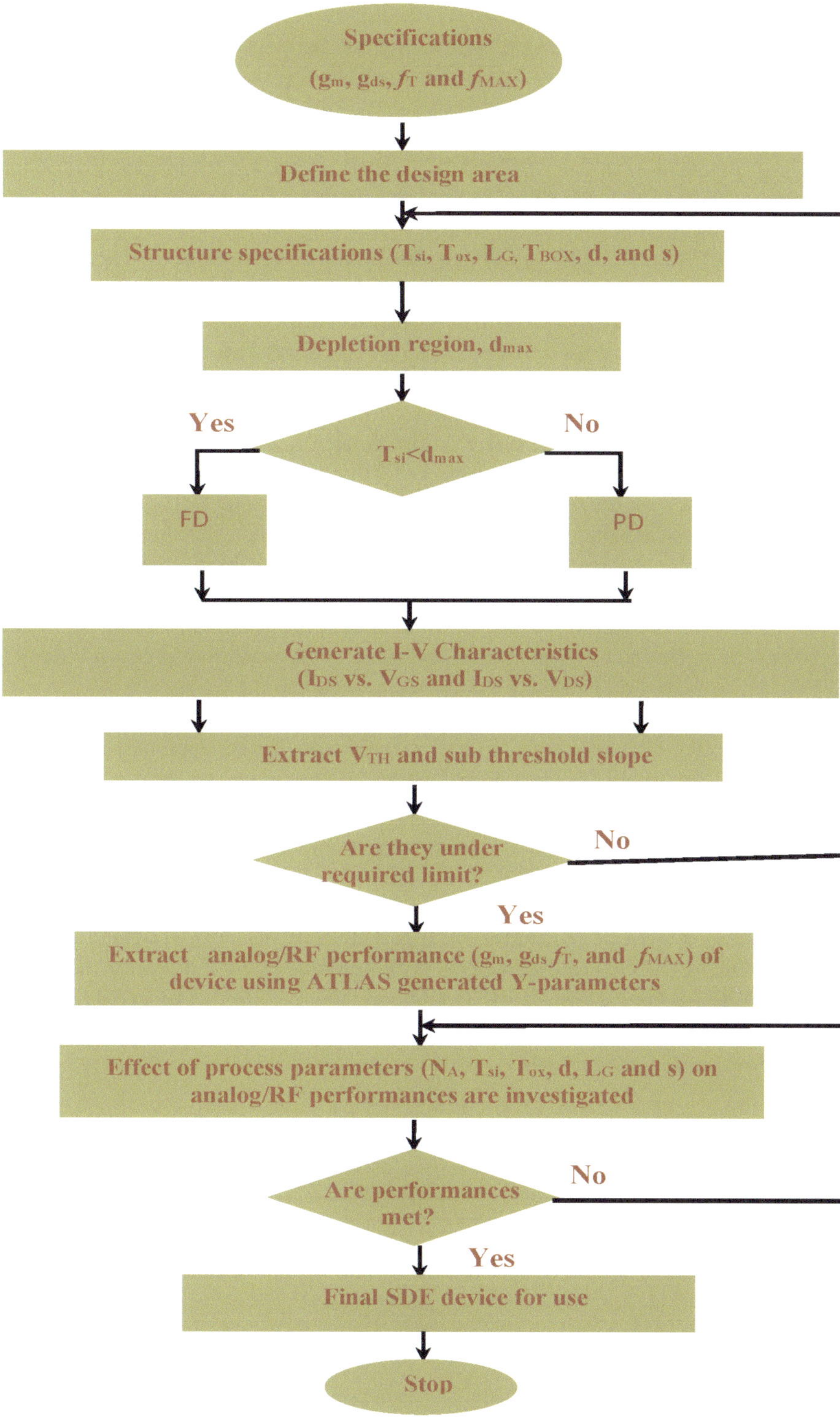

Figure 9.
Device design flow chart.

Mid gap material gate work function w of 4.62 eV corresponding to the High Performance (HP) 32 nm node logic technology has been used in this work. After optimization T_{si} was found as $\cong 0.3 \times L_G$ for maintaining full depletion and minimizing SCEs and extra source/drain series resistance problem. In order to keep the device almost free from SCEs, the natural length γ as discussed in above Section 4.1 was found $\cong 15$. A device is free from SCEs if the effective gate length of gate-underlap device is larger than five to ten times the natural length γ [18] that's why in this design gate length L_G has been chosen as 90 nm.

According to ITRS-2011 [20], road map 32 nm high performances (HP) logic technology, analog technology is always running behind 2–3 decades with logic, therefore, this gate length L_G as 90 nm is found realistic.

A significantly thinner buried oxide (BOX) ($\cong$20–50 nm) could somewhat improve DIBL and SCEs, but this would come at the cost of reduced gate control and enhanced junction capacitance. Furthermore, a thin BOX leads to the serious increase in the parasitic capacitance as well as the degradation on the g_m and question most of the benefits of the SOI concept [30], therefore, BOX thickness T_{BOX} was taken 400 nm in this design [30]. For minimum parasitic effects and higher value of I_{on}, g_m, f_T and f_{MAX}, spacer s $\cong 0.8 \times L_G$ was taken with fixed doping gradient (d = 5 nm/decade). It is because of lower d (d < 5 nm/decade) corresponds to an abrupt phenomenon, which is hard to realize in actual fabrication scenario whereas higher d (d > 5 nm/decade) would cause significant fluctuation in the dopant underneath the channel resulting in undesirable effects. The optimized design parameters used in simulation of gate-underlap SOI MOSFETs are tabulated in **Table 2**.

Source/drain region was modeled by a gate-underlap design, using Gaussian source/drain profile with lateral straggle σ, across a spacer s defined by the distance from the start of the profile to the edge of the gate. The lateral straggle σ, is directly related to s and d the inverse source/drain doping gradient in nm/decade evaluated at the gate edge, by $\sigma = \sqrt{2sd/\ln(10)}$ as discussed in Section 3.2. In underlap design, S/D profiles are designed with s/σ =2.8 for optimal analog performance as well as extension region s doping does not significantly extend beyond the gate edge [17].

The simulations were performed with drift diffusion model (DD), field dependent mobility model (FLDMOB) and Lambardi CVT (known as CVT mobility model) model using ATLAS device simulator. The DD model, which is used to calculate the

Parameter	Value
Silicon film thickness, T_{si}	$0.3 \times L_G$ nm
Gate oxide thickness, T_{ox}	3 nm
Gate length, L_G	90 nm
Device width, W	64 μm
Buried oxide thickness, T_{BOX}	400 nm
Threshold voltage of device, V_{TH}	0.26 V
Spacer, s	$0.8 \times L_G$ nm
Channel doping concentration, N_A	1×10^{16} cm^{-3}
S/D doping concentration, N_D	5×10^{20} cm^{-3}

Table 2.
Device design parameters.

potential distribution and current density inside the device and FLDMOB model incorporates the effect of velocity saturation, whereas CVT mobility model calculate the effects of mobility, dependent on doping N_A, temperature T, vertical field (E↑) and parallel field (E//) [34]. Quantum effects have not been incorporated due to operation in low moderate inversion region and higher silicon film thickness (T_{si} > 10 nm). Source/drain contact was made Ohmic. Gate–underlap channel design can be fabricated using a dual spacer process technology as discussed in [33]. Finally, the device has been designed using the steps given in flow chart (see **Figure 9**) and simulated structure of gate-underlap device is shown in **Figure 10**. The doping gradient of the designed device along cut line in the channel region is shown in **Figure 11**

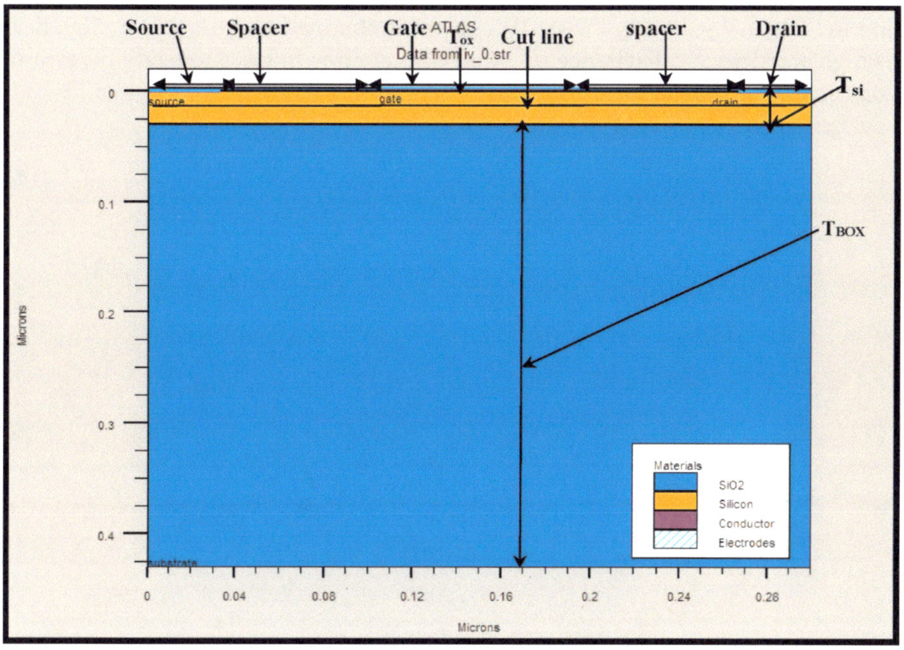

Figure 10.
ATLAS simulated gate-underlap device structure.

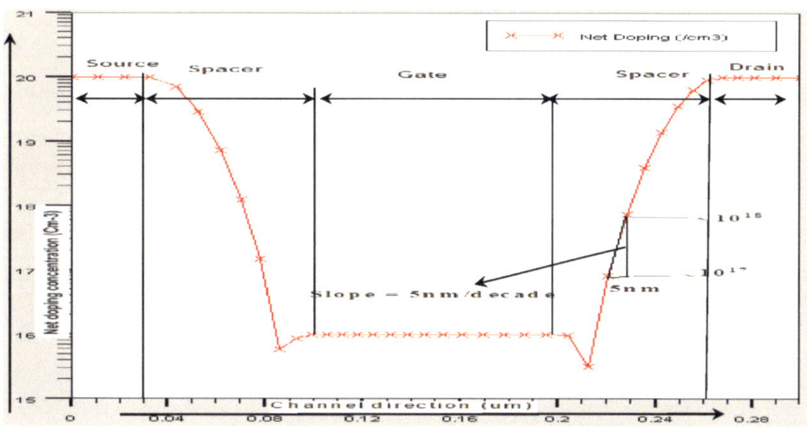

Figure 11.
Illustration of a gate-underlap S/D doping profile (x-x) with d = 5 nm/dec s ≅ 0.8 × L_G.

and found a reasonable agreement of the simulated value with the theory which has been explained in the previous section.

5.2 DC-IV characteristics

Figure 12(a) shows the plot of I_{DS} versus V_{GS} of designed device. As our region of interest for operation is low moderate inversion i.e. $V_{GS} - V_{TH} \leq 100$ mV [4]. It is because of SDE-region optimization is particularly advantageous in the weak and low moderate inversion region ($V_{GS} - V_{TH} \leq 100$ mV) as this region the current flow is mainly due to the diffusion of carriers [4]. In this design in low moderate inversion region (at $V_{OD} = V_{GS} - V_{TH} = 90$ mV) current is obtained $\cong$15 μA/μm. In SDE design g_m is high and device capacitance (C_{gs} and C_{gd}) are low in low moderate inversion region resulting in high f_T and intrinsic gain A_v. It can be seen from **Figure 12(b)** that the designed device is free from kink-effects.

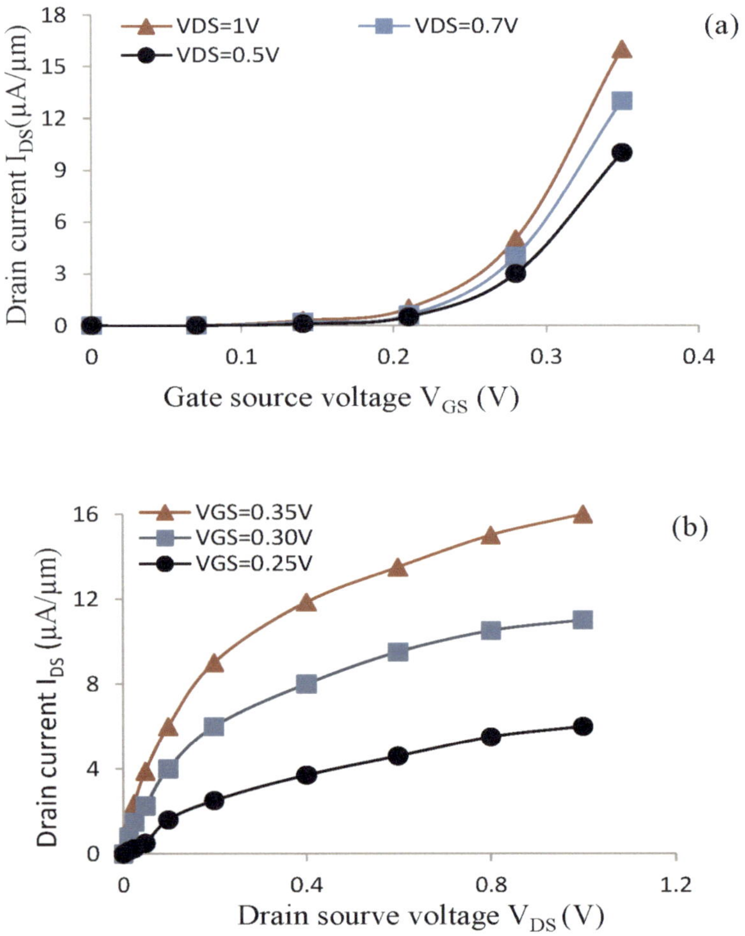

Figure 12.
Plot of curve (a) $I_{DS} - V_{GS}$ (b) $I_{DS} - V_{DS}$.

5.3 Effects of process parameters on threshold voltage

Device physical structure and channel engineering are two options to improve the performance of device for high frequency low power applications. As discussed in Chapter 2 the V_{TH} of FD-SOI MOSFETs is sensitive to process parameters. The sensitivity of V_{TH} vs. process parameters such as T_{si}, T_{ox}, s and L_G at fixed doping gradient d = 5 nm/decade are shown in **Figure 13(a)–(f)**.

The V_{TH} variation with channel doping at different T_{si} can be seen from **Figure 13 (a)**. The V_{TH} is changing with T_{si} and the channel doping concentration. It is found that V_{TH} is very sensitive to T_{si} for higher N_A which is not desirable. With increasing the N_A, V_{TH} increases which is quite acceptable according to (15). With increasing N_A, the Fermi potential increases; also channel depletion charge increases; it takes more effort to deplete the whole channel. That's why the V_{TH} increases with increasing the N_A. In SOI devices, the T_{si} determines the amount of source/drain charge sharing and the coupling between back and the front gates. Thicker T_{si} with the same N_A increases the depletion charge, resulting in increases the V_{TH} as shown in **Figure 13(b)**. The desired V_{TH} ($\cong$ 0.26 V) is found at $T_{si} \cong 0.3 \times L_G$ with N_A = 1E16 cm^{-3} (almost undoped).

Figure 13(c) shows the V_{TH} variation with oxide thickness T_{ox} at different T_{si}. From the simulation results, it can be seen that with increasing T_{ox} from 2 nm to 4 nm, the V_{TH} also increases. This is due to fact that with increasing T_{ox} the gate capacitance decreases, and the gate has less control to the channel. V_{TH} variation with spacer s can be seen from **Figure 13(d)**, it is clear from the result, that V_{TH} increases with increase in s. It is due to the fact that longer spacer results in increase in depletion charges as a result V_{TH} increases and appropriate V_{TH} can be found with s $\cong 0.8 \times L_G$.

Furthermore, to study the impact of gate length L_G on device, the L_G is varied from 30 nm to 120 nm. As shown in **Figure 13(e)**, with L_G increasing V_{TH} decreases, until L_G = 90 nm, after that V_{TH} increases again with increase in L_G. This is because of at lower value of L_G (L_G < 90 nm) geometry goes to down, dopant increase and there is a dopant fluctuation inside the channel (non-uniformly charge distribution inside the channel) as a result V_{TH} is increased whereas for higher value of L_G (L_G > 90 nm) there is a higher depletion charge in channel region hence again V_{TH} value is increased.

As discussed above, varying N_A, T_{si}, T_{ox}, s and L_G, V_{TH} can be controlled. But for geometry < 100 nm, it is very difficult to control the channel doping. For example, when the device channel length is 100 nm, silicon film thickness 50 nm, channel width 1 μm, with this small geometry, the total volume is 5E−15 cm^{-3} [12]. In order to get the doping concentration 2E−17cm^{-3} the total doping doze is 1000 atoms, which is a very small number. It is difficult to get the accurate channel doping concentration in the real technology. Therefore, random fluctuation of the doping concentration can lead to variations of $V_{TH.}$

Another attractive way to adjust the V_{TH} is the gate electrode work function engineering [6]. By varying gate work function w, the flat band voltage can be changed hence V_{TH} can also be changed. **Figure 13(f)** shows the V_{TH} variation with gate work function w. It can be seen from figure that w increases V_{TH} increase and it is well in conformity, with formulation given in (16). In order to maintain good SCEs performance and obtained proper value of V_{TH}, the gate work functions of nMOSFETs and pMOSFETs must be close to those of n+ and p+ doped material [35]. Therefore, in this work mid gap gate work function is preferred (w = 4.62 eV).

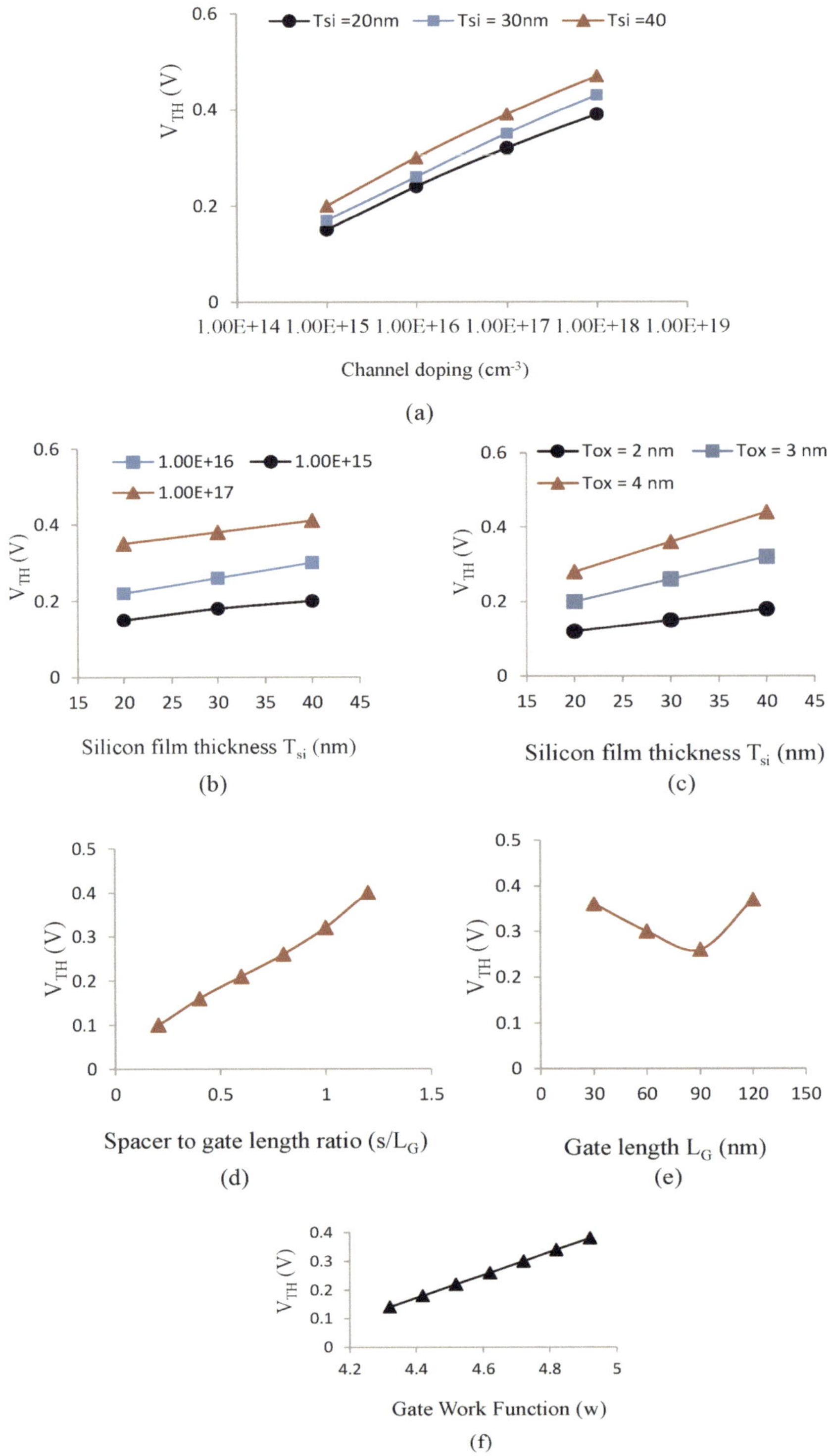

Figure 13.
Threshold voltage variations with (a) channel doping at different silicon film thickness (b) silicon film thickness at different channel doping concentration level (c) silicon film thickness with different gate oxide thickness (d) spacer (e) gate length and (f) gate work function, at V_{DS} = 1 V and V_{OD} = 90 mV.

For low power application need low V_{TH}. The SDE engineering is more beneficial in low moderate inversion region. For robust application of the device in GHz range applications supply voltage should be minimum three times that of V_{TH}. Through simulation, V_{TH} of the device was found as ≅0.26 V. At this value of V_{TH}, optimized process parameters of the device (s, T_{si}, T_{ox}, L_G, N_A and w at fixed d = 5 nm/decade) are tabulated in **Table 2**.

5.4 Effects of process parameters on intrinsic gain A_v, f_T and f_{MAX}

In this section for GHz range applications device has been verified in-terms of intrinsic gain A_v, f_T and f_{MAX} with varying process parameters and the obtained results are shown in **Figures 14–17**.

From **Figure 14(a)**, it can be seen that with increasing T_{si}, g_m decreases whereas g_{ds} increases. It is due to the fact that higher V_{TH} results in reduced I_{on} therefore, the

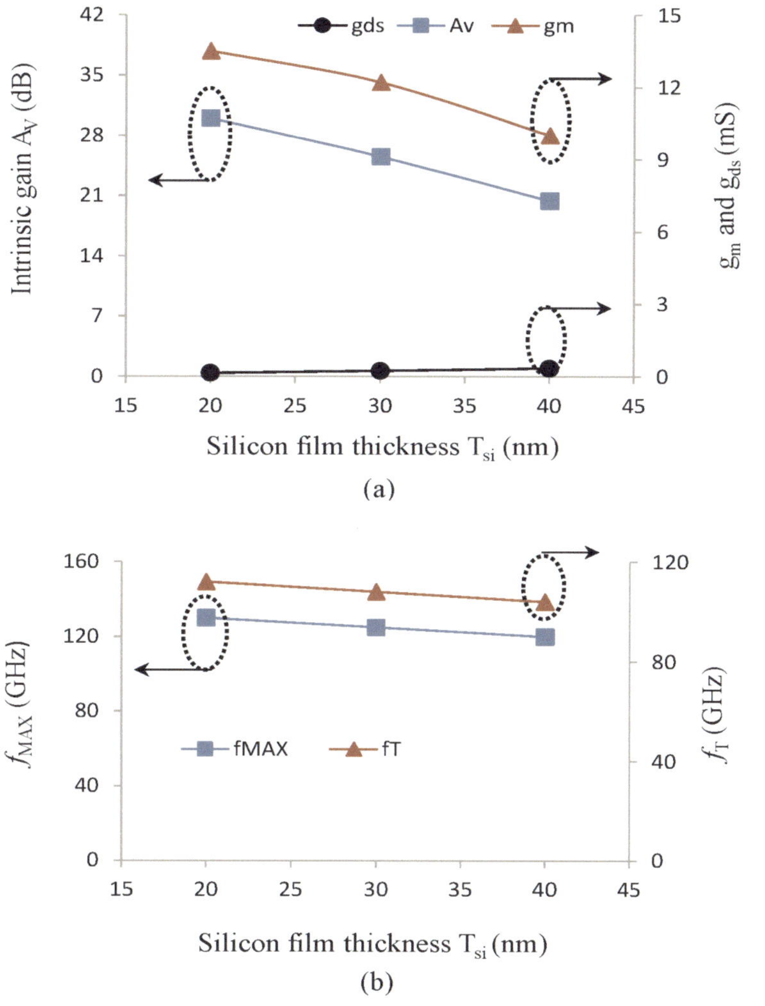

Figure 14.
Variations of (a) transconductance g_m, output conductance g_{ds}, intrinsic gain A_v and (b) transit-time frequency f_T and maximum oscillatory frequency f_{MAX} vs. silicon film thickness T_{si}.

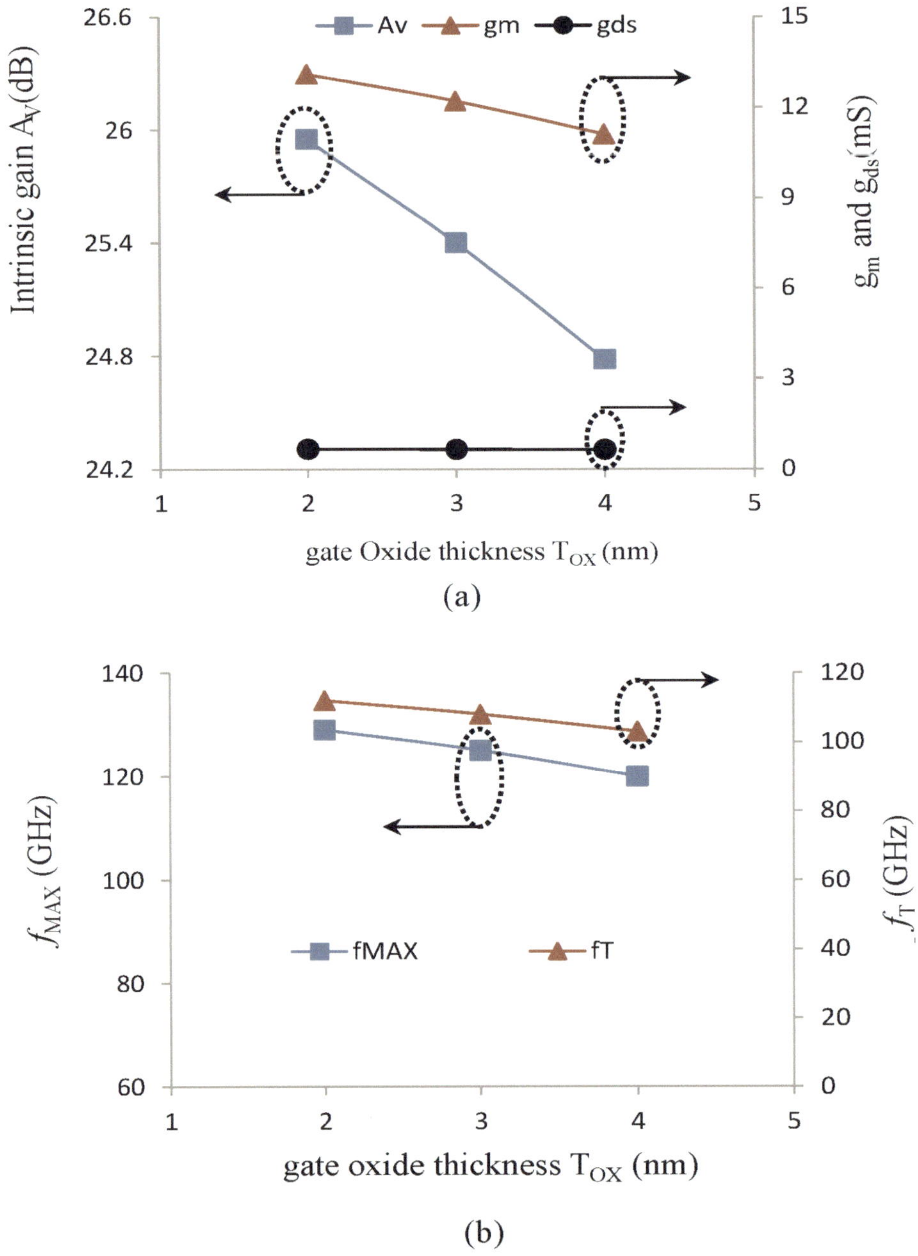

Figure 15.
Variations of (a) transconductance g_m, output conductance g_{ds}, intrinsic gain A_v (b) transient-time frequency f_T and maximum oscillatory frequency f_{MAX} vs. oxide thickness T_{ox}.

value of g_m is reduced remarkably whereas g_{ds} is increased significantly as a result the intrinsic gain A_v of the device suffer significantly, (see **Figure 14(a)**). Furthermore, for higher T_{si} thickness, g_m is reduced noticeably due to less control of SCEs and also the device parasitic capacitance increases as a result f_T and f_{MAX} decreases as shown in **Figure 14(b)**. This is also due to the fact that f_T and f_{MAX} is inversely proportional to parasitic drain series resistance/drain series resistance, and directly proportional to g_m [12].

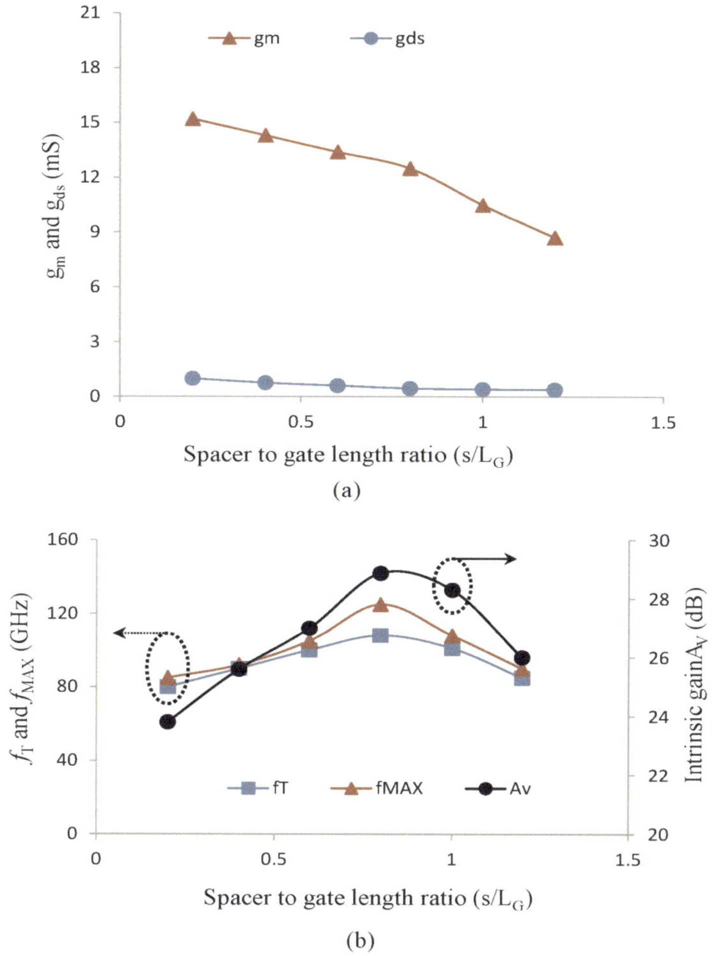

Figure 16.
Variations of (a) transconductance and output conductance (b) intrinsic gain, transit- time frequency and maximum oscillatory frequency vs. spacer.

As shown in **Figure 15(a)**, g_m decreases moderately whereas g_{ds} decreases marginally as a result A_v decreases with increase in T_{ox}. This is due to the fact that with increase in gate oxide thickness, the gate capacitance decreases, and the gate has less control to the channel. In order to invert the channel, the V_{TH} will increase to compensate it. Therefore, I_{on} decreases as a result g_m decreases. In addition, with increase in T_{ox} thickness gate to channel capacitance is lowered down as result g_m is decreased, as a consequence of that f_T and f_{MAX} are decreased with increasing T_{ox} thickness (see **Figure 15(b)**).

With increase in s, the effective gate length (L_{eff}) increases so carriers do not travel at the saturation velocity over virtually the whole of channel, as a result "on" current I_{on} decreases, therefore, value of transconductance (g_m) decreases remarkably as shown in **Figure 16(a)**. However, by increasing s, the electric field at the gate edge towards drain has been reduced, which results in reduced values of output conductance (g_{ds}) sharply (see **Figure 16(a)**) as a result A_v (g_m/g_{ds}) improves significantly (see **Figure 16(b)**). Furthermore, **Figure 16(b)** represents the variation of f_T and f_{MAX} with variable s/L_G with doping gradient d = 5 nm/decade at low $V_{OD} \leq 90$ mV.

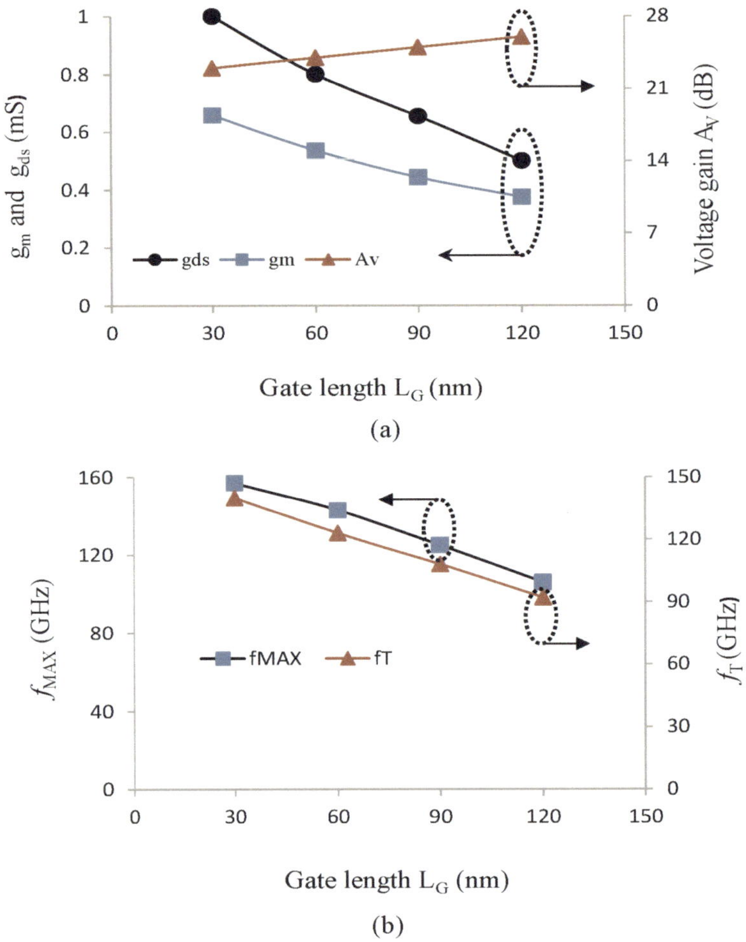

Figure 17.
Variations of (a) transconductance, output conductance and intrinsic gain (b) transit-time frequency and maximum oscillatory frequency vs. gate length L_G (device width W = 64 μm).

f_T increases with increase in s due to marginally decrease in g_m and significantly higher decrease in overall parasitic capacitances up to s $\cong$0.8 × L_G., and after s > 0.8 × L_G the value of g_m decrease significantly and source/drain series resistance contributed their effects noticeable, hence f_T start to decrease with increase in s/L_G ratio. f_{MAX} follows the f_T trend and improves with s in the range ≤ 0.8 × L_G. From the **Figure 16(b)**, it can be seen that f_T and f_{MAX} as well as intrinsic gain A_v achieved maximum value at optimal spacer $s \cong$ 0.8 × L_G.

From **Figure 17(a)**, it can be concluded that with increase in gate length L_G, g_m and g_{ds} are reduces and intrinsic gain A_v improves slightly. This is supported by g_m and g_{ds} curves as shown in **Figure 17(a)**. In addition, f_T and f_{MAX} decreases with increase in L_G. This is due to the fact that increased parasitic and decreased g_m values (see **Figure 17(b)**).

From the above discussions it has been found that V_{TH} of FD-SOI MOSFETs is sensitive to process parameters and device is free from kink effects. The process parameters of the device like T_{si}, T_{ox}, and s are optimized and found approximately as 0.3 × L_G, 3 nm and 0.8 × L_G, respectively where L_G = 90 nm at V_{OD} = 90 mV.

The width of designed device W was taken as 64 m for optimum noise performance [33]. The source/drain (SDE) region engineered technique shows the potential of the present technology for low power/low-voltage GHz applications. In the next chapter designed device is modeled and performances for high frequency application have been carried out.

6. Conclusions

In this chapter different device design approaches such as thin body, graded channel, halo doped, multiple-gate and SDE have been explained. Among them, SDE was found very effective approach to combat SCEs as well as for improved high frequency analog/RF performance. The basic device physics involved in the designing of the SDE has been discussed. Furthermore, the effects of process parameters like T_{si}, T_{ox}, L_G and s, on threshold voltage V_{TH}, RF performance matrices like A_v, f_T and f_{MAX}, have been carried out in low moderate inversion region, targeting for low power and high frequency applications. Above results suggested that sub 100 nm FD-SOI MOSFETs will be strong contenders for analog/RF applications in modern lucrative wireless communications market.

Author details

Indra Vijay Singh
Scientist-C, ICMR Project, MGMIHS, Navi Mumbai, India

*Address all correspondence to: vijaysinghmgm35@gmail.com

References

[1] Dennard R. The impact of Dennard's scalling theory. IEEE Solid-State Circuit Society News. 2007;**12**(1)

[2] Belestra F, Cristoloveanu S, Benachair M, Birani J, Elewa T. Doublegate silicon-on-insulator transistor with volume inversion: A new device with greatly enhanced performance. IEEE Electron Device Letter. 1987;**8**:410-412

[3] Kranti A, Chung TM, Flandre D, Raskin JP. Laterally asymmetric channel engineering in fully depleted SOI-MOSFETs for high performance analog applications. Solid State Electronics. 2004;**48**:947-959

[4] Kranti A, Armstrong GA. Engineering source/drain extension regions in nanoscale double gate SOI MOSFETs: Analytical model and design consideration. Solid-State Electronics. 2006;**50**(3):437-447

[5] Choi Y-K, Asano K, Lindert N, Subramanian V. Ultrathin-body SOI MOSFET for deep-sub-tenth micron era. IEEE Electron Device Letter. 2000;**21**(5)

[6] Deshpande HV, Cheng B, Woo JCS. Channel engineering for analog device design in deep submicron CMOS technology for system on chip applications. IEEE Transactions on Electron Devices. 2002;**49**(9):1558-1565

[7] Narasimhulu K, Sharma DK, Rao VR. Impact of lateral asymmetric channel doping on deep submicron mixed-signal device and circuit performance. IEEE Transactions on Electron Devices. 2003; **50**(12):2481-2489

[8] Ke W, Han X, Xu B, Liu X, XinanWang TZ, Han R, et al. Source/drain series resistances of nano-scale ultra-thin-body SOI MOSFETs with undoped or very-low-doped channel regions. Semiconductor Science and Technology. 2006;**21**:1416-1421

[9] Cerdeira A, Alemán MA, Pavanello MA, Martino JA, Vancaillie L, Flandre D. Advantages of the graded-channel SOI FD-MOSFET for application as a quasi-linear resistor. IEEE Transactions on Electron Devices. 2005;**52**(5)

[10] Pavanello MA, Martino JA, Dessard V, Flandre D. An asymmetric channel SOI- nMOSFET for reducing parasitic effects and improving output characteristics. Electrochemical and Solid-State Letters. 2000;**3**:50-52

[11] Pavanello MA, Martino JA, Flandre D. Comparison of analog performance in conventiional and graded-channel fully depleted SOI MOSFETS. In: XV Int. Conf. On Microelectronics and Packaging. 2000. pp. 67-71

[12] Najeeb-ud-din V, Rao R, Vasi J, Woo JC. Channel engineering for high speed sub −1.0 V power supply deep sub micron CMOS. In: Technical Digest, Symposium on VLSI Technology. 1999. pp. 69-70

[13] Taur Y, Wann CH, Frank DJ. 25 nm CMOS design consideration. IEDM Technical Digest. 1998:789-792

[14] Vanacailly L, Kilchytska V. Influence of halo implantation on analog performance and comparison between bulk, partially-depleted and fully depleted MOSFETs. In: IEEE SOI Conference. 2002. pp. 161-163

[15] Hakim N, Rao VR, Vasi J, Woo JC. Small signal characterization of thin film single halo SOI MOSFET for mixed mode analog and digital applications. In: 16th

Int. Conf. On VLSI Design. 2003. pp. 110-115

[16] Park J-T, Colinge J-P. Multiple-Gate SOI MOSFETs: Device design guidelines. IEEE Transactions on Electron Devices. 2002;**49**(12)

[17] Kranti A, Armstrong GA. Comparative analysis of nano-scale MOS device architectures for RF applications. Semiconductor Science and Technology. 2007;**22**:481-491

[18] Colinge JP. Multi-gate SOI MOSFETs. Microelectronic Engineering. 2007;**84**:2071-2076

[19] Alam MS, Kranti A, Armstrong GA. Inter-modulation non-linearity investigation of nano-scale gate-underlap double gate MOSFETs. In: Proc. of European Silicon-on-Insulator International Conference (EUROSOI 2010), from 25–27, Grenoble, France. 2010. pp. 57-58

[20] International technology roadmap for semiconductor. 2011 edition. Available from: http://public.itrs.net

[21] Cheng B, Rao VR, Woo JCS. Exploration of velocity overshoot in a high performance deep submicron-0.1-um SOI MOSFET with asymmetric channel profile. IEEE Electron Device Letter. 1999;**20**(10):538-540

[22] Depande HV, Cheng B, Woo JCS. Analog device design for low power mixed mode application in deep submicron CMOS technology. IEEE Electron Device Letter. 2001;**22**: 588-590

[23] Shin H, Lee S. A 0.1 um asymmetric halo by large-angle tilt implant MOSFET for high performance and reliability. IEEE Transactions on Electron Devices. 1999;**46**(4):820-822

[24] Chen Q, Agrawal B, Meindl JD. A comprehensive analytical suthreshold swing (S) model for double gate SOI MOSFETs. IEEE Transactions on Electron Devices. 2002;**49**:1086-1090

[25] Chen Q, Herrel EM, Meindl JD. A physical short-channel threshold voltage for undoped symmetric double gate MOSFETs. IEEE Transactions on Electron Devices. 2003;**50**:1631-1637

[26] Kranti A, Chung TM, Flared D, Raskin JP. Analysis of quasi double gate method for performance prediction of deep submicron double gate SOI MOSFETs. Semiconductor Science Technology. 2005;**20**:423-429

[27] Cristoloveanu S, Ernst T, Munteanut D, Ouisse T. Ultimate MOSFETs on single and double gate or ground plane. International Journal of High Speed Electronics and Systems. 2000;**10**: 217-230

[28] Jaju V, Dalal V. Silicon-on-insulator technology. In: EE 530, Advances in MOSFETs. Springer; 2004

[29] Auth CP, Plummer JD. Scaling theory for cylindrical, fully-depleted, surrounding-gate MOSFET's. IEEE Electron Device Letter. 1997;**18**(2):74-76

[30] Ernst T, Muteanu D, Cristoloveanu S, Ouisse T, HeFene N, Horiguchi S, et al. Ultimately thin SOI MOSFETs: Special characteristics and mechanisms. In: Proceedings of the IEEE International SOI Conference. 1999. pp. 92-93

[31] Yan RH, Ourmazd A, Lee KF. Scaling the Si MOSFET: From bulk to SOI to bulk. IEEE Transactions on Electron Devices. 1992;**397**:1704-1710

[32] Suzuki K, Tanaka T, Tosaka Y, Horie H, Arimoto Y. Scaling theory for double

gate SOI MOSFETs. IEEE Transactions on Electron Devices. 1993;**40**:326-329

[33] Alam MS, Kranti A, Armstrong GA. Investigation of gate-underlap design on linearity of operational transconductance amplifier. In: Proc. of the WCECS Computer Science Vol II; October 20–22; USA. 2010

[34] ATLAS-2012 Device Simulator. Available from: http://www.silvaco.com

[35] Kim K, Fossum JG, Chuang CT. Process/physics based threshold voltage model for nano-scaled double-gate devices. International Journal of Electronics. 2004;**91**:139-148